International Conference on Renewable Energy and Environment Engineering (REEE 2018)

E3S Web of Conferences Volume 80 (2019)

Paris, France
29 - 31 October 2018

Editor:

Wenzhi Cui

ISBN: 978-1-5108-7991-1

Printed from e-media with permission by:

Curran Associates, Inc.
57 Morehouse Lane
Red Hook, NY 12571

Some format issues inherent in the e-media version may also appear in this print version.

Printed by Curran Associates, Inc. (2019)

For additional information, please contact EDP Sciences – Web of Conferences
at the address below.

EDP Sciences – Web of Conferences
17, Avenue du Hoggar
Parc d'Activité de Courtabœuf
BP 112
F-91944 Les Ulis Cedex A
France

Phone: +33 (0) 1 69 18 75 75
Fax: +33 (0) 1 69 28 84 91

contact-edps@webofconferences.org

Additional copies of this publication are available from:

Curran Associates, Inc.
57 Morehouse Lane
Red Hook, NY 12571 USA
Phone: 845-758-0400
Fax: 845-758-2633
Email: curran@proceedings.com
Web: www.proceedings.com

TABLE OF CONTENTS

Preface

2018 The International Conference on Renewable Energy and Environment Engineering (REEE 2018) was successfully held in Paris, France on October 29th to 31st.

The proceedings are the result of REEE 2018, which presents 24 accepted articles selected from more than 60 papers submitted to the conference from 20 countries and districts including Portugal, Canada, South Africa, Poland, Japan, Philippines, Estonia, China, Morocco, Romania, Algeria, Ecuador, Russia, Spain, Rwanda, Malaysia, etc. All of the papers were subjected to peer-review by the program technical committee members and international reviewers. All the papers are presented orally in two parallel sessions.

We had three very interesting keynote speakers Prof. Nidia Caetano, School of Engineering (ISEP), Polytechnic Institute of Porto (IPP), Portugal; Prof. Carlos Felgueiras CIETI/ISEP, Portugal and Prof. Saad Mekhilef, University of Malaya, Malaysia.

On behalf of the organizing committee, I would like to express my sincere gratitude to guest editors, program technical committee members and international reviewers, who provided constructive criticism and stimulated comments and suggestions to the authors. It's a tough work to select papers from the submissions, and without their hard work the proceedings will not see the light. Besides, we also want to thank the chairs, who are dedicated to make sure to run the conference smoothly, the keynote speakers and invited speakers who shared with us their precious research findings.

Finally, I hope the readers will find the proceedings interesting and valuable. Hope to see you in the REEE conference in the near future.

Yours Sincerely,

Prof. Saad Mekhilef
Program Chair
University of Malaya, Malaysia

Statement of Peer review

In submitting conference proceedings to E3S Web of Conferences, I certify to the Publisher that I adhere to the **Policy on Publishing Integrity** of the journal in order to safeguard good scientific practice in publishing.

1. All articles have been subjected to peer review administered by the proceedings editors.
2. Reviews have been conducted by expert referees, who have been requested to provide unbiased and constructive comments aimed, whenever possible, at improving the work.
3. Proceedings editors have taken all reasonable steps to ensure the quality of the materials they publish and their decision to accept or reject a paper for publication has been based only on the merits of the work and the relevance to the journal.

Title, date and place of the conference

2018 International Conference on Renewable Energy and Environment Engineering

Paris, France | October 29-31, 2018

Proceedings editor(s):

Wenzhi Cui
Professor, Chongqing University, China

Date and editor's signature

17, avenue du Hoggar - PA de Courtabœuf – BP 112 - 91944 Les Ulis Cedex A (France)
Tél. : 33 (0)1 69 18 75 75 - Fax : 33(0)1 69 07 45 17 – www.edpsciences.org

Ecological briquettes from sunflower seed husk

Cosmin Spirchez[1], *Aurel* Lunguleasa[1], and *Cătălin* Croitoru[2]

[1]Transilvania University of Brasov, Wood Processing and Design Wooden Product Department, 29 Street Eroilor, Brasov, Romania
[2]Transilvania University of Brasov, Engineering Materials Department, 29 Street Eroilor, Brasov, Romania

Abstract. The paper aims to use the waste of sun-flower hulls in order to obtain some lignocellulosic briquettes as renewable solid fuels. First, the sunflower seed husks were ground, after that they were sorted and only the fractions corresponding to the briquettes and pellets were used. After obtaining the stable briquettes there were determined the physical and caloric properties of briquettes manufactured from sunflower husks, as potential alternative biofuels. The moisture content, ash content and density of the briquettes have been determined and correlated with their calorific properties, namely the calorific value and calorific density. The experimental results have indicated that, even if the self-adhesion of sunflower seed hulls is less than that of wood sawdust, they have a better behaviour in humid environment. Good results make it possible to use these briquettes for both household heating systems and thermal power plants.

1 Introduction

Nowadays there is an impinging potential crisis of fossil fuels, as research shows a shortage for oil and coal in the next 50 – 60 years [1,2]. Wood that traditionally was used as energy source since ancient times is currently under regulation, as most of countries from the European Union are concerned with the preserving of the natural forest biotopes [2]. Due to these regulations imposed at national level, even inferior quality wood (obtained from branches and boughs), becomes less available [3,4].

Biomass is a renewable energetic source, because it increases from one year to another, it is widely spread world-wide and presents low costs in comparison to the fossil fuels, the biomass resources, from which fuel material is produced may include wood and wooden wastes, agricultural cereals and wastes resulted from their production, aquatic biomass and algae [12,15].

Wood industry by-products, such as sawdust, have been successfully used for the manufacturing of briquettes with satisfactory caloric properties, but nowadays a great deal of attention are focused on other type of biomass-derived materials, such as fruit seeds, peels and so forth [3-8].

By appropriately modifying the properties of these biomass-derived materials, for example by thermal treatment, they could be transformed in superior fuels such as briquettes and pellets.

Oleaginous seed husks have attracted attention lately due to their generation as byproduct in considerable amounts in the oil production industry [9] and to their potential use as high caloric-output materials, aided by the trace amounts of fatty polyesters.

A sun-flower seeds processing unit has a current capacity of 400 tons in 24 hours, from which it results approximately 100 tons of seed shelling.

Although the sun-flower seed shelling are wastes of the main production process (Fig 1), it is necessary to manage them, because they can lead to blockages in the storage facility of the factories.

The main uses of sunflower seed shelling are in the briquettes and pellets field, and in the lignocellulose composites. They are reported to combust with minimal emissions of volatile organic compounds in the atmosphere, thus making them useful as an alternative ecologic fuel.

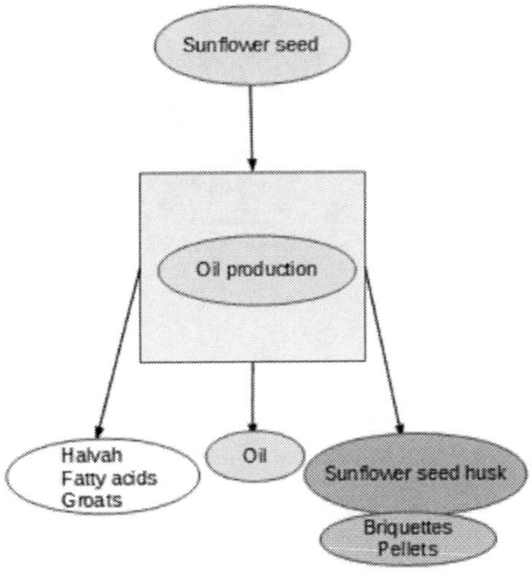

Fig. 1. Sunflower oil production flow

2 Materials and methods

For determining the effective density of the briquettes, 12 pieces of pellets have been chosen and machined to obtain regular cylindrical specimens, usually by sanding ends with disc machine. The effective density is usually determined as a ratio between mass and volume. The relation in Eq.1 (considering the cylindrical form of briquettes) had been used in order to obtain effective density of briquettes:

$$\rho_{ef} = \frac{4m}{\pi \cdot d^2 \cdot l} \; (g/cm^3) \qquad (1)$$

Where:
m is mass of briquettes, in g;
d- diameter of briquettes, in mm;
l- length of briquettes, in mm.

From these briquettes, some samples with smaller lengths were extracted in order to be used for compressive strength. To determine the breakout endurance of the materials, a universal mechanical testing instrument was used, where there were installed two compression plates, as see in Fig 2.

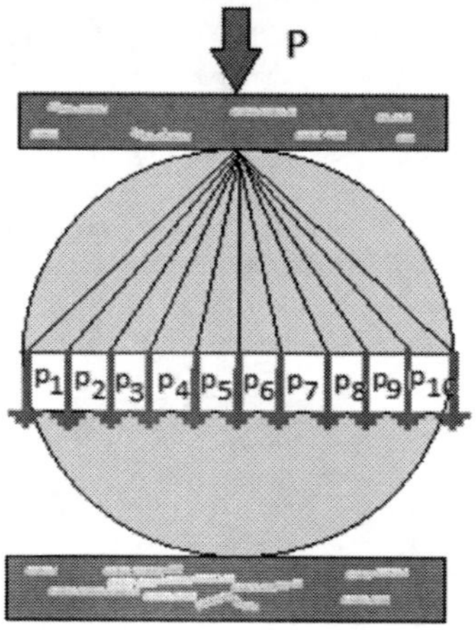

Fig. 2. Determining the compression strength of briquettes

The compressive strength of briquettes is determined as a ratio between force and area of analyzed surface (Eq.2).

$$\sigma_c = \frac{F}{A} \; (N/mm^2) \qquad (2)$$

where:
σ_c is the compressive strength, in N/mm^2;
F- Maximum force of briquettes crash, in N;
A –area of compressive crash surface, in mm^2.

Fig 2 show that all stresses are concentrated in the middle zone of briquettes, namely the broken surface is given by diameter and length of these samples, that form a specifically area of breakage. Taking in consideration the above aspects, the compressive strength will be determined with the next relation:

$$\sigma_c = \frac{F}{d \cdot l} \quad [N/mm^2]$$

σ_c is the compressive strength, in N/mm^2;
F- Maximum force of briquettes crash, in N;
d- Diameter of briquettes, in mm;

where:
σ_c is the compressive strength, in N/mm^2;
F- Maximum force of briquettes crash, in N;
d- Diameter of briquettes, in mm;
l- length of briquettes samples, in mm.

The installation used for determining the calorific value of the wooden biomass was the explosive burning (bomb) calorimeter type XRY-1C, produced by Shanghai Changji Geological Instrument Co., from China (Fig. 3).

Before testing the calibration of the calorimetric bomb was performed with benzoic acid with a known value of calorific value (26 463 kJ/kg). With this value of calorific value, the calorimetric coefficient noted with *k* of calorimeter will be obtained, using the normal procedure of determination and equation of CV determination (Eq.3).

$$CV = \frac{k \cdot (t_f - t_i)}{m} - \sum_{i=1}^{n} q_i \quad [kJ/kg] \qquad (3)$$

where:
CV is calorific value, in kJ/kg;
k- calorific coefficient of calorimeter, in kJ/Celsius;
t_f –final temperature, in Celsius;
t_i-initial temperature, in Celsius;
m-mass of sample, in kg;
$\sum_{i=1}^{n} q_i$ is supplementary heat release in time of testing, in kJ/kg.

Supplementary heat is given by burning nickel-chromium and cotton wires and also by burning of HNO_3 (usually 40 J).

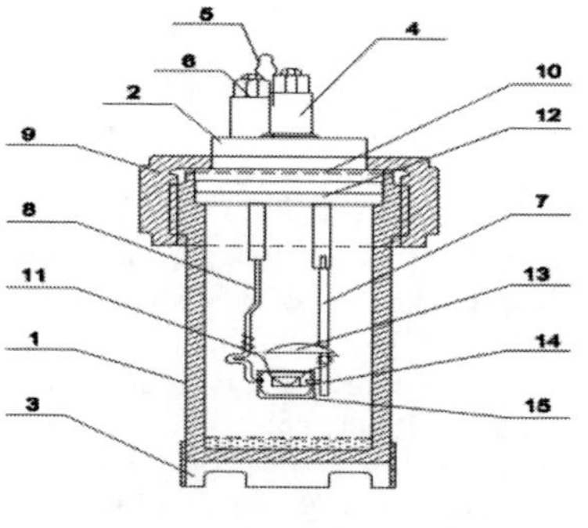

Fig. 3. Section view through calorimetric bomb

The inferior caloric power of the sunflower husk pellets is determined based on the superior caloric power, expressed in Eq.4:

LCV= HCV– 6 (MC + 9h) [kJ/kg] (4)

where:

LCV- inferior caloric power, kJ/kg

MC- wood moisture content of sample, kJ/kg

h- hydrogen content of the sample, estimated at 3.6%, for the lignocellulosic materials.

The testing sample (1) is tied to the cotton thread (2) and is put in the calorimeter chamber (3). The copper nickel thread (4) is tied to the sample and the cotton thread, after which the protection covers (5) is attached. The calorimeter chamber is connected to the calorimetric bomb cover 6 through two electrodes (7) and (8), which continue with the electrical threads for calorimetric bombs coupling (9) and (10). By threading the cover, the bomb (11) is coupled to fitting (12) to the oxygen supplying unit, generating a total pressure of 30 atmospheres.

In fig. 4 the description of the process to determine the calorific power is presented, as there were presented by the computer software. The preparation of the test facility requires a series of outdoor activities, but also the introduction of data on the computer interface (such as sample weight, calorific value, etc.)

The test procedure has three main and different steps [11, 18]:

The first stage ("fore") has as purpose the determination of temperature variations of water in the calorimetric recipient, due to the heat exchange with the exterior before burning. During this period, usually for 5 minutes, the temperature is determined at specified intervals with the precision thermocouple.

The last reading of the temperature from the first period represents in fact the first temperature in the "main" period. The values of the temperature registered in this period are six. After registering the 6th value the flaming/ignition of the material takes place and its reading on the menu bar ("burning time").

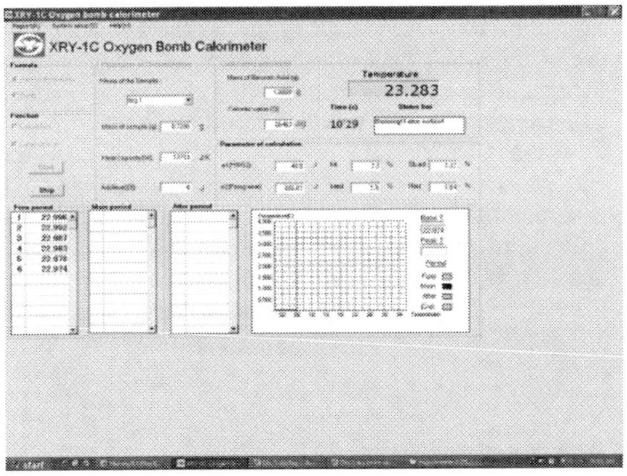

Fig. 4. Description of the process to determine the calorific power

The main period ("main") begins by burning (flaming) the sample and has as consequence the increase of water temperature in the calorimetric recipient, due to the burning of the pellet and heat delivery to container. The values of the temperature are indicated at one-minute periods. The values registered during this period vary according to the burning time of the fuel material in the calorimetric bomb. The number of registered temperature values may vary between 19 – 42 during this period. These periods are usually equaled wit time of this period in minutes.

The final period ("after") has as purpose the determination of the average water temperature in the calorimetric recipient, due to the heat exchange with the exterior vessel after burning.

Identical to the first stage, the temperature is shown at half minute periods, for 4-5 mutes and on average there are registered 8 – 10 values of the temperature variation. The last temperature value is the final temperature value and is recorded by the computer software. After that, the computer software displays the end of the test by word and position "end " by software.

For the superior calorific power, the value obtained was of 19256 kJ/kg, while for the inferior calorific power, the average value obtained was 18647 kJ/kg.

To determine the ash content of the seed husks, the general method of standardized determination was used (ASTM D2866-11, 2012). According to this method, the milled and dried material (dried to 0% moisture content) is calcined at a temperature of 750°C in a laboratory calibrated oven, for a period of at least 3 hours (Fig. 5).

The pellets calcination is made on a metallic melting crucible resistant to elevated temperature of 750 Celsius degrees, and the weighting was made on an analytical balance with a 3-decimal precision.

When determining the ash content, it will be taken into consideration that the sample is completely dried and the cleaned and empty melting pot weight, as in Eq.5 [12, 13].

$$A_c = \frac{m_{a+c} - m_c}{m_{s+c} - m_c} \cdot 100 \quad [\%] \qquad (5)$$

where:

m_{a+c} -mass of ash, considering the crucible, in g;

m_{s+c}- mas of sample considering the crucible, in g;

m_c- mass of empty crucible, in g.

Fig. 5. Oven for determining the ash content

3 Results and discution

The average mean of the sunflower briquettes density is 0.9934 (g/cm^3) (Fig.6). Large variations could be due to the differences in compactness.

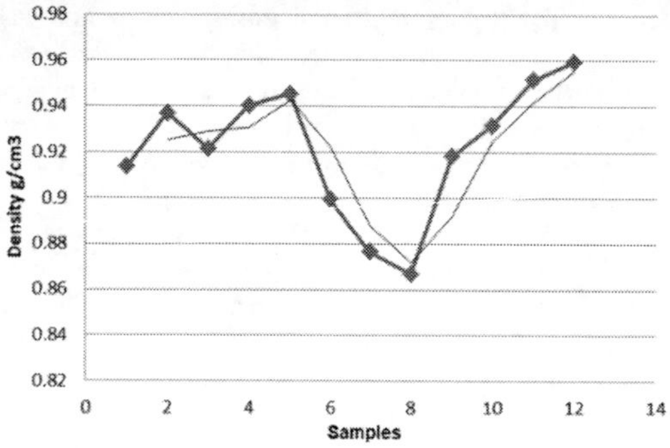

Fig. 6. The density of the briquettes

Fig.7 depicts the observed breakout endurance of briquettes, for the twelve samples taken into analysis.

The average breakout resistance or crash strength for the briquettes is 0.3-0.6 MPa.

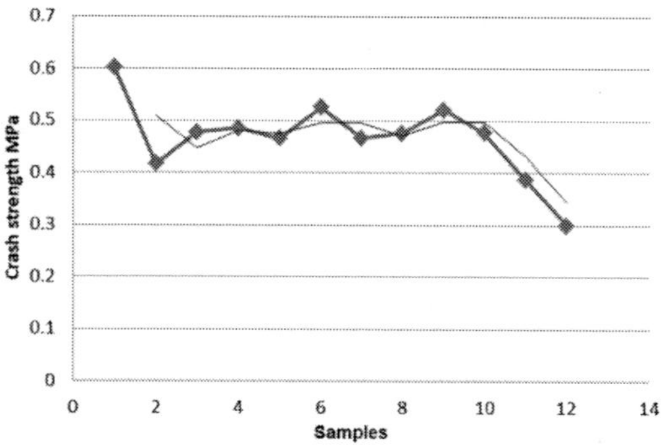

Fig. 7. Compressive strength of briquettes

In fig. 8 presented the variation of ash content, active carbon and volatile substances.

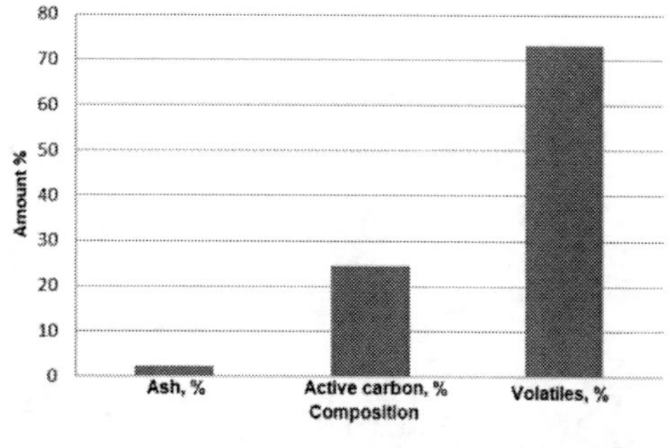

Fig. 8. Ash content, active carbon and volatile substances in the briquettes

The amount of ash contained is about 2.4%, the amount of active carbon is 22.7%, and the amount of volatile is 72% as determined in other research [14].

4 Conclusions

The sun-flower seeds shelling are a waste from sun-flower oil factory. These huge wastes may be used successfully in briquettes, after their grinding, sorting and drying without using other additives. From this point of view these obtained briquettes are some natural fuel materials that can be used successfully for obtaining heat in stoves or furnaces.

The results presented in the paper have indicated that the obtained briquettes could in principle successfully replace the wooden sawdust briquettes or even the fire wood, due to the good density and the very good caloric power. The ash content of 2.4% is like some of European wooden species. The burning of briquettes obtained from sun-flower husk produce a clean energy, used for stoves or furnaces.

This kind of briquettes has not significantly influence the general properties of briquettes but it makes a major contribution to their combustion properties because of low moisture content. This is one of the main reasons (apart from economic arguments) why many briquette producers are going to do next step namely to change raw materials from wood to agricultural waste.

The crushing strength values obtained in the paper should cover the real crushing strengths, even for the most restrictive standards. Procedure of determination of breaking strength is very simple and elucidates the compactness and consistency of briquettes, thus improving the quality and properties of these products. Consequently this study has practical importance for producers and customers of briquettes.

Biomass is one of the forms of renewable sources which may be converted in solid, liquid and gaseous energetic fuel and which may generate energy as heat by its burning, as well as electrical energy by conversion processes.

Presently the biomass contributes with approximately 12% to the production of primary energy in the world, and in the countries in developing process it covers 40-50% of the necessary of energy. The Biomass is the alternative source that, according to contributed with 7% from the energy produced in the world. Presently, the use of renewable fuel materials such as wooden wastes for producing biofuels increases the chances of biomass in the availability level on the energetic market.

Biomass is environmentally friendly and a neutral energy against the emissions of carbon dioxide. The carbon dioxide is absorbed by the plants during the growing process and forms a closed circuit, because the quantity of carbon dioxide which was absorbed by the plants during the growing process will be equal to the one which was eliminated during the complete burning process.

Acknowledgement

We wish to acknowledge stuff of Transilvania University of Brasov, Project nr. 8063/2.07.2018 (The program of quality evaluation of briquettes from sunflower seed husk).

References

1. T. Banta, R. De Leona, Parametric study of rice husk torrefaction for the development of sustainable

solid fuel International Journal of Smart Grid and Clean Energy Vol **7**, No 3, pp. 207-217 (2018)

2. A.V. Bridgwater, Review of Fast Pyrolysis of Biomass and Product Upgrading, Biomass bioenergy, vol. **38**, pp. 68-94 (2012)
3. J.P. Boutin, G. Gervasoni, R. Help, K. Seyboth, P. Lamers, M. Ratton, et al Altrnative energy sources in Transition Countries. The case of Bio-energy in Ukraine, Environmental Engineering and Management Journal, vol. **6,** nr.1, pp.3-11 (2007)
4. A. Demirbas Resource Facilities and Biomass Conversion Processing for Fuels and Chemicals, Energy Conversion Management, vol. **42,** nr. 11, pp. 1357-1378, (2011)
5. A.M. Garcia, B.M.J. Barcia, J.A. Hernandez Preparation of active Carbon from a Comercial Holm-Oak Charcoal: Study of Micro and Meso-porosity, Journal Wood Science and Technology, vol. **37,** nr. 5, pp. 499-509 (2008)
6. D. Gavrilescu Energy from biomass in pulp and paper, Environmental Engineering and Management Journal, vol. **7(5)**, pp.537-546, (2008)
7. M. Gavrilescu Biomass Power for Energy and Sustainable Development, Environmental Engineering and Management Journal, vol. **7(5)**, pp.617-640, (2008)
8. N. Kaliyan, R.V. Morey Factors Affecting Strength and Durability of Densified Biomass Product, Biomass Bioenerg, vol. **33**, nr. 3, pp. 379-359, (2009)
9. T. Kuhlman, V. Diego, E. Koomen Exploring the potential of reed as a bioenergy crop in the Netherlands, Biomass Bioenergy, vol. **55**, pp.41-52, (2013)
10. J. Lako, J. Hancsok, T. Yuzhakova, G. Marton, A. Utsai, A. Redey, Biomass and Bioenergy , vol. **40,** pp. 63-73, (2012)
11. A. Lunguleasa Compaction coefficient of wooden briquettes used as renewable fuel, Environmental Engineering and Management Journal, vol.**10,** nr. 9, pp.1263-1268, (2011)
12. C. Okello, S. Pindozzi, S. Faugno, L. Boccia Bioenergy potential of agricultural and forest residues in Uganda, Biomass Bioenergy, vol. **56**, pp. 515-525, (2013)
13. N. A. Pambudi, K. Itaoka, A. Chapman, N. D. Hoa, N. Yamakawa Biomass energy in Japan: Current status and future potential, International Journal of Smart Grid and Clean Energy, vol. **6**, no. 2, pp. 119-126 (2017)
14. B. Popescu, L. Şenilă, C. Vărăticeanu, G. Şimon Cellulosic bioethanol from sunflower seed hulls-a renewable energy source, Studia UBB Ambientium, vol. **1-2**, pp.105-110, (2013)
15. J. Swithenbank, Q. Chen, X. Zhang, V. Sharifi, M. Pourkashamiani Wood would burn, Biomass and bioenergy, vol.**3**, (2011)
16. J. Tanchuling, R. De Leon Solid fuel characterization of torrefied coconut shells in an oxidative environment, International Journal of Smart Grid and Clean Energy, vol. **7**, no. 2, pp. 145-151 (2018)
17. O. Teuch A. Hofeanuer, F. Troger, J. From Basic properties of specific wood based materials carbonised in a nitrogen atmosphere, Wood Science Technology Journal, vol.**38,** nr.3, pp.323-333, (2004)
18. V.K. Verna, S. Bram, J. de Rucky Small Scale Biomass Systems: Standards, Quality, Labeling and Market Driving Factors-An Outlook Biomass Bioenerg, vol.**33**, nr. 10, pp. 1393-1402, (2009)

E3S Web of Conferences **80**, 01002 (2019)
REEE 2018
https://doi.org/10.1051/e3sconf/20198001002

Daily global solar radiation estimation based on air temperature: case of study south of Algeria

Razika Ihaddadene[1], *Nabila* Ihaddadene[1], *Mohamed El Hacen Ould Ahmedou Bemba* Jed[1] and *Amaury De* Souza[2]

[1]Departement of Mechanics, M^{ed} Boudiaf University M'sila, Algeria
[2]Federal University of Mato-Grosso do Sul, C. P. 549, 79070-900 Campo Grande, MS, Brazil

Abstract. Global solar radiation is needed for the analysis and scaling of solar conversion systems; however, global measurements of solar radiation are not available in all Algerian cities. The use of empirical models using an accessible parameter is a solution to this problem. In this study, seven empirical Models namely Hargreaves and Samani, Chen, M.F. Li, H.Li, Bristow and Campbell, Okonkwo and Abraha Savage have been employed to estimate daily average global solar radiation on the horizontal surface. These models use extreme temperatures (minimum and maximum). They were applied to three south Algerian sites (Biskra, Ghardaia, and Tamanrasset). The analyzed data were provided by the NASA site and cover four years (2001-2004). The validation of the models for predicting daily global solar radiation was done using four statistical parameters (R^2, MBE, RMSE, and RPE). The results show that Bristow and Compbell Model shows better performance than the other models in all sites. A new model is proposed for each site. The results show that this later is the best one compared with the seven models analyzed. Therefore, the developed model can be suggested to estimate daily global solar radiation using only extreme air temperatures in south Algeria.

1 Introduction

Due to its geographic location, Algeria has the largest solar field in the world. The sunshine duration of virtually all the national territory exceeds 2000 hours annually and can reach 3900 hours in High lands and in Sahara. The daily-obtained energy on a horizontal surface of 1 m^2 is about 5 kWh over the major part of the national territory, or about 1700 $kWh/m^2/year$ in the north and 2263 $kWh/m^2/year$ in the south of the country [1]. The measurement stations of various components of solar radiation are scarce or almost non-existent on all the national territory. While general meteorological parameters (pressure, temperature (max and min), sunshine duration, humidity, wind speed .., etc.) are available at Meteorology National Office stations. Therefore, it is interesting to develop mathematical models to describe the various components of solar radiation as a function of these parameters.

Numerous models have been established to estimate the amount of global solar radiation on a horizontal surface using various climatic parameters, such as sunshine duration[2,3,4], cloud cover[2,5,6], humidity[7], maximum and minimum ambient temperatures[8,9,10,11,12,13], wind speed…etc.

Several studies have shown that performing models use sunshine duration data followed by those using cloudiness and temperature data [14-15] respectively. However, sunshine duration data are not as readily available as air temperature data at standard meteorological stations, while, the temperature is an interesting and easily accessible meteorological parameter. Therefore, it is important to elaborate models that estimate solar radiation based on air temperature only. Due to the availability of daily maximum and minimum air temperature, several empirical models have been proposed to estimate solar radiation from these variables. Hargreaves and Samani[8] proposed the first empirical model to estimate solar radiation using only maximum and minimum air temperatures. Another model was proposed by Bristow and Campbell[11]. This later is a simple model with different configurations, in which; global solar radiation G is an exponential function of temperature difference. These two models have been a subject of several modifications, which comprise adding further parameters or introducing other meteorological variables.

Based on the work of Hargreaves and Samani, Allen suggested employing a self-calibrating model to estimate mean monthly global solar radiation [10]. Chen et al. [11] have refined the Hargreaves and Samani model by adding another coefficient. Samani et al. [14] have replaced the constant of Hargreaves and Samani model by a polynomial function of second order of temperature difference.

In addition, H Li et al. [15] proposed a modification of the Chen model by assuming that the coefficient of Hargreaves and Samani model is a linear function of the average temperature. Danatelli and Campbell [16] have modified the Bristow and Campbell model by taking into account two corrective functions of the mean and minimum air temperatures. Goodin et al. [17] have refined the Bristow and Campbell model by adding a G_0-term meant to act as a scaling factor allowing temperature difference (T_{max}-T_{min}) to accommodate a greater range of G values.

Similarly, other modifications were made to Hargreaves and Samani model by Weiss et al. [18] Abraha et al. [19], and Almorox et al. [1]. Li et al. [20] proposed another form of the evolution of global solar radiation as a function of the minimum and maximum temperatures. Okonkwo et al [21] have proposed a model for estimating global solar radiation received on a horizontal surface in Minna using a polynomial function of the third order of temperature ratio T_{min}/T_{max}.

The global solar radiation measurements are not generally operational in all cities of Algeria, due to the high cost of acquisition, maintenance, calibration and technical complexities. To overcome this problem, the use of empirical models is necessary. In this study, an analysis of the possibility of reliably estimating daily global solar radiation, from air temperature data in three cities of south Algeria; Biskra, Ghardaia, and Tamanrasset is proposed. The objectives of this study are, firstly to evaluate the accuracy and applicability of seven models that use the extraterrestrial radiation and air temperatures (maximum and minimum) to the three cities, and then select the most accurate model which estimates the daily global solar radiation for each city. Secondarily, to develop a simple new model that gives a good estimation of daily global solar radiation from the maximum and minimum daily temperatures of the same regions cited above. Then the performance of the developed model is validated for each city by comparing it, with measured data using different statistical tests.

2 Materials and methods

2.1 Collection data

In order to evaluate the applicability and accuracy of the seven models used for computing the daily global solar radiation on a horizontal surface, the geographical and meteorological data of the three south Algerian cities (Biskra, Ghardaia, and Tamanrasset) are used. The latitude, longitude and elevation for each location are presented in Table1.

Table 1. The cities terrestrial coordinates.

Cities	Latitude	Longitude	Altitude(m)
Biskra	34°41'	5°44'	87.
Ghardaia	32°29'	3°41'	461
Tamanrasset	22°47'	5°31'	1400

The daily global solar radiation data and temperature data (maximum and minimum temperatures) of each city, varied from the first January until the 31st December of the years of 2001, 2002, 2003 and 2004. These data were provided by the NASA web site.

2.2 Models description

Seven models are used in this study; they describe the solar radiation as a function of extraterrestrial radiation and maximum and minimum temperature. Extraterrestrial radiation is given as follows:

$$G_0 = \left(\frac{24x60}{\pi}\right) I_{sc} d_r \left[\cos(\varphi)\cos(\delta)\sin(\omega_s) + \omega_s \sin(\varphi)\sin(\delta)\right] \quad (1)$$

where:

I_{sc} : the solar constant (equal to 1367 W/m²),

d_r :the relative earth-sun distance, it is given by the following formula:

$$d_r = 1 + 0.033\cos\left(\frac{2\pi}{365} j\right) \quad (2)$$

j : the number of day,

φ : the latitude of the location under consideration (in radian),

δ : the solar declination (expressed in radian), it is given as :

$$\delta = 0.4093\sin\left(\frac{2\pi}{365}(248 + j)\right) \quad (3)$$

ω_s : the sunset hour angle, it is given by the following expression :

$$\omega_s = arc\left[-\tan(\varphi)\tan(\delta)\right] \quad (4)$$

Brief descriptions of the models employed in this analysis are as follows:

2.2.1 Hargeaves and Samani Model (Model 1)

This model estimates the global solar radiation using only maximum and minimum temperature; Hargeaves and Samani have developed the following correlation [8]:

$$G / G_0 = a_1 + (T_{max} - T_{min})^{0.5} \quad (5)$$

where:

G : is the monthly average daily global solar radiation [kWh/m²],

G_0 : is the monthly average daily extraterrestrial solar radiation [kWh/m²],

a_1 : is an empirical coefficient.

2.2.2 Chen Model (Model 2)

Chen et al. [11] have modified the model of Hargreaves and Samani by introducing an empirical coefficient, b_2 as given by the following equation:

$$G / G_0 = a_2 + (T_{max} - T_{min})^{0.5} + b_2 \quad (6)$$

2.2.3 M. F. Li Model (Model 3)

Li et al. [20] suggested the following correlation to estimate the daily global solar radiation as a function of minimum temperature (T_{min}) and maximum temperature

(T_{max}) as follows:

$$G/G_0 = a_3 + b_3 T_{max} + c_3 T_{min} \quad (7)$$

Where, a_3, b_3 and c_3 are empirical coefficients.

2.2.4 H. Li Model (Model 4)

Li et al. [15] proposed a model to estimate monthly average global solar radiation for all climatic regions in China. This model is the result of changes made to the models of Hargeaves & Samani and Chen. It is a function of the temperature difference ΔT (T_{max} -T_{min}) and average air temperature T_a. Li et al. model [10] is given as follows:

$$G/G_0 = (a_4 + b_4 T_a)\Delta T^{0.5} + c_4 \quad (8)$$

Where:

T_a: is the monthly average air temperature (°C), defined as follows:

$$T_a = (T_{max} + T_{min})/2 \quad (8.a)$$

a_4, b_4 and c_4 are empirical coefficients.

2.2.5 Bristow and Compbell Model (Model 5)

Bristow and Compbell [9] proposed a model to estimate the daily global solar radiation as an exponential function of the air temperature difference (ΔT = T_{max} -T_{min}) as noted below:

$$G/G_0 = a_5\left[1 - \exp(-b_5 \Delta T^{c_5})\right] \quad (9)$$

a_5, b_5 and c_5 are the empirical coefficients of this model.

2.2.6 Okonkwo Model (Model 6)

Okonkwo et al. [21] proposed a new model for estimating global solar radiation in Minna (Nigeria) by using a quadratic model based on the temperature ratio (T_{min}/T_{max}). It is given by the following formula:

$$G/G_0 = a_6 + b_6(T_{min}/T_{max}) + c_6(T_{min}/T_{max})^2 + d_6(T_{min}/T_{max})^3 \quad (10)$$

a_6, b_6 ,c_6 and d_6 are the empirical coefficients of this model.

2.2.7 Abraha Savage Model (Model 7)

Abraha Savage [22] proposed a new model for estimating global solar radiation in Cortez (Colorado, USA), Davis (California, USA), Padova (Italy), Rothamsted (UK), Wageningen (Netherlands), Pretoria (South Africa), Griffith (Australia) by using an exponential model based on the temperature difference (T_{max} - T_{min}). It is given by the following formula:

$$G/G_0 = 0.75 \times \left[1 - \exp(-a_7 \Delta T^2 / \Delta T_m)\right] \quad (11)$$

a_7 is the empirical coefficient of this model and ΔT_m is the medium value of temperature difference.

2.3 Models performance validation

To illustrate the agreement between predicted and measured values of global solar radiation of the seven models, four types of judgment criterion were used; the coefficient of determination (R^2), the root mean square error (RMSE), the mean bias error (MBE), and the mean percentage error (MPE). They are frequently used to evaluate the models' performance [23,24,25].

3 Results and discussion

In this study, seven empirical models mentioned above namely; Model 1, Model 2, Model 3, Model 4, Model 5, Model 6 and Model 7 are compared in order to determine their suitability to estimate the global solar radiation received on a horizontal surface at three south Algerian cities (Biskra, Ghardaia and Tamanrasset). The empirical coefficients of those models obtained using Matlab software from the daily global solar radiation data of the cited sites are regrouped in Table 2.

Table 2. Emperical coefficients values of the models in selected cities.

Models	a	b	c	d
Mod1				
Bis	0,143			
Gha	0,146			
Tam	0,146			
Mod2				
Bis	0,006	6 0,494		
Gha	0,007	0,494		
Tam	0,002	0,628		
Mod3				
Bis	0,374	0,007	-0,0003	
Gha	0,401	0,008	-0,0009	
Tam	0,601	0,003	-0,002	
Mod4				
Bis	-0,0006	0,002	0,442	
Gha	0,007	0,002	0,448	
Tam	0,018	0,0001	0,570	
Mod5				
Bis	0,606	0,0007	3 ,21	
Gha	0,649	0,006	2,26	
Tam	8,72	0,063	0,08	
Mod6				
Bis	0,539	-0,088	0,503	-0,067
Gha	0,529	0,327	-0,614	0,919
Tam	0,674	0,063	-0,056	-0,080
Mod7				
Bis	0,065			
Gha	0,073			
Tam	0,102			

After determining the coefficients of each model, the global solar radiation value for each day following each model is calculated. A comparison between the computed and measured values of daily global solar radiation was performed using the statistical indicators cited above (R^2, RMSE, MBE, and MPE). The values of these statistical indicators are regrouped in Table 3. As the statistical analysis shows Model 7 (Abraha Savage) is the least representative model of all models in the sites of Biskra and Ghardaia. It represents the lowest coefficient of determination (0,8314 for Biskra site and 0,8760 for Ghardaia as noted in Table 3). For Tamanraset site the least model is Model1.

Table 3. Statistical indicators values of models studied.

Models	R^2	MBE (Wh/m²/D)	RSME (Wh/m²/D)	MPE (%)
Mod1				
Bis	0,8669	-0,17	0,832	1,738
Gha	0,8944	-0,15	0,789	1,691
Tam	0,7694	-0,07	0,569	-0,022
Mod2				
Bis	0,9314	-0,17	0,748	2,597
Gha	0,9404	-0,15	0,728	2,150
Tam	0,8883	-0,01	0,393	0,472
Mod3				
Bis	0,9265	-0,06	0,581	1,696
Gha	0,9284	-0,06	0,573	1,401
Tam	0,8874	-0,01	0,393	0,479
Mod4				
Bis	0,9256	-0,06	0,585	1,712
Gha	0,9276	-0,06	0,576	1.412
Tam	0,8874	-0,01	0,393	0,479
Mod5				
Bis	0,9359	-0,16	0,719	2,479
Gha	0,9406	-0,15	0,706	2,093
Tam	0,8884	-0,02	0,395	0,434
Mod6				
Bis	0,9172	-0,11	0,660	2,165
Gha	0,9224	-0,09	0,621	1,688
Tam	0,8860	0,20	0,444	4,102
Mod7				
Bis	0,8314	0,01	0,857	4,836
Gha	0,8760	-0,03	0,767	3,646
Tam	0,7709	0,09	0,567	2,681
New Mod				
Bis	0,9413	0,001	0,502	0,784
Gha	0,9521	0,.001	0,449	0,617
Tam	0,8914	0,004	0,384	0,559

All the sites of Biskra, Ghardaia and Tamanrasset follow the same order of the models, Model5 is the best one followed by Model2, then Model3 and Model4 and finally Model6.

It is observed from all these Models that the ratio of global solar radiation to the extraterrestrial radiation (G/G_0) depends of the maximum temperature (T_{max}) and the minimum temperature (T_{min}). The variation of daily global solar radiation (G) at Biskra, Ghardaia and Tamanrasset cities according to the daily extraterrestrial radiation (G_0) follows a linear law. The determination coefficients of these correlations are of 0.9413, 0.9512 and 0.8914 respectively. In addition, the evolution of daily global solar radiation received on horizontal surface at these cities is proportional to the temperature sum ($T_{max}+T_{min}$) for Biskra(R^2=0.622), Ghardaia(R^2=0.625) and Tamanrasset(R^2=0.609).

Based on these results, the following models aiming to estimate the daily global solar radiation received on horizontal surface at Biskra, Ghardaia and Tamanrasset sites, can be proposed. They are presented as a function of daily extreme temperatures (T_{max}, T_{min}) and daily extraterrestrial radiation. They are given as follows for Biskra, Ghardaia and Tamanrasset sites:

$$G = -1{,}578 + 0{,}8.3G_0 + 0.000701\,(T_{max} + T_{min}) \quad (12)$$

$$G = -1{,}864 + 0.8877G_0 - 0{,}00431(T_{max} + T_{min}) \quad (13)$$

$$G = -0{,}3566 + 0{,}7633\,G_0 - 0{,}01274\,(T_{max} + T_{min}) \quad (14)$$

For a better representation of our results a comparison between the monthly global solar radiations measured and calculated using this new model and model5 was performed for the three sites (Biskra, Ghardaia and Tamanrasset)(Figure 1). As seen, the calculated values using the new model fit well to the measured ones.

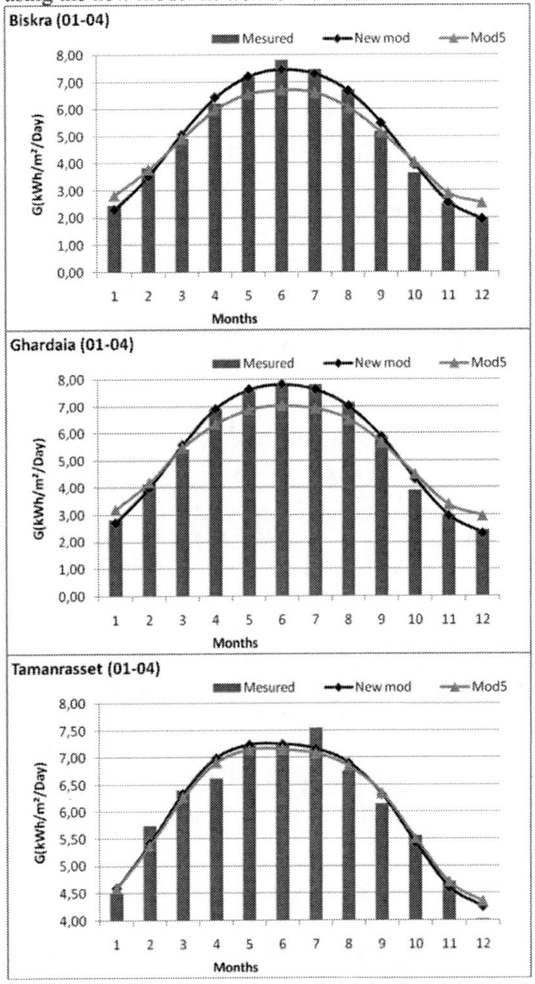

Fig. 1. Comparaison between monthly global solar radiation measured and calculated using the new model and Model5 for all sites..

For Biskra site, the new model for daily global solar radiations has a coefficient of determination R^2 of 0,9413, which is greater than that of the best model Model 5 with a value of 0,9359 as seen in Table 3. It presents a value of 0,7840 (%) for MPE, a value of 0,502 (kWh/m²/Day) for RMSE and a value of 0.001 (kWh/m²/Day) for MBE, which are less than those found for Model 5 (2,479 (%) for MPE,0.719 for RMSE and -0.16 (kWh/m²/Day) for MBE). For Ghardaia site, the proposed model has a coefficient of determination R^2 of 0.9521, which is greater than that of the best model Model 5 with a value of 0,9406. It presents a value of 0,617(%) for MPE, a value of 0,449(kWh/m²/Day) for RMSE and a value of 0.001 (kWh/m²/Day) for MBE, which are less than those found for Model 5 (2,093(%) for MPE, 0.706 for RMSE and -0.15 (kWh/m²/Day) for MBE).

For Tamanrasset site, the proposed model has a coefficient of determination R^2 of 0,8914 which is greater than that of the best model Model 5 with a value of 0.8884. It presents a value of 0.559(%) for MPE, a value of 0.384(kWh/m²/Day) for RMSE and a value of -0.004 (kWh/m²/Day) for MBE, which are less than those found for Model 5 (0.395 for RMSE and -0.02 (W/m²) for MBE).

4 Conclusion

Our model describes daily global solar radiations as a linear function of extraterrestrial radiation and the sum of the maximum and minimum temperature. The coefficients of this model depend on the chosen site. It was applicable for three sites of Algeria. It is the best one in the series of the models studied. It allows determining the evolution of daily global solar radiation using only the extreme temperature values (maximum and minimum) in the south of Algeria.

References

1. M. R. Yaich, A. Bouhanik, S. M. A. Bekkouche, A. Malek, and T. Benouaz, Energ. Convrs. Manage, **82**, (2014)
2. F. Besharat, A. A. Dehghan, A. R. Faghih, Renew Sust. Energ. Rev, 21, (2013)
3. A. Das, J-k. Park, J-h. Park, J. Atmos. Sol-Terr. Phy,**134**, (2015)
4. M. EL-Metwally, J. Atmos. Sol-Terr. Phy, **67**, (2005)
5. L. Supit, RR. V. Kappel, Sol. Energy, **63**, (1998)
6. V. Badescu, Theor. Appl. Climato, **72**, (2002)
7. PE. Thornton, SW. Running, Agric. For. Meteorol, **93**, (1999)
8. G. H. Hargeaves and Z. A. Samani, J. Irrig. Drain. Div, **108**(3), (1982)
9. K. L. Bristow, and G. S. Campbell, Agric. For. Meteorol, **31**(2), (1984)
10. R. Allen, J. Hydrol. Eng, 2, (1997)
11. R. Chen, K. Eric, J. Yang, S. Li, and W. Zhao, Energ. Convrs. Manage, **45**(11-12), (2004)
12. R. Meenal, P. G. Boazina, A. I. Selvakumar , *Indian J. Sci Technol,* **9**(46), (2016)
13. J.D.S Valdineny, R.D.S. Claudio, A.J. José Aust. *J. Crop Sci,* **10**(7), (2016)
14. Z. Samani, J. Irrig. Drain. Div *Eng,* **126**(4), (2000)
15. H. Li, F. Cao, X. Wang and W. Ma, The Sci World J, **3**: (2014)
16. M. Donatell, G.S. Campbell, Proceeding sof the fifth European society of agronomy congress, Nitra, Slovak Republic. (1998)
17. J. D. G. Goodin, M. S. Hutchinson, R. L. Vanderlip, M. C. Knapp, Agron. J, **91**, (1999)
18. A. Weiss, C. J. Hays, Q.Hu, W. E. Easterling, Agron. J, **93**, (2001)
19. M. G. Abraha, M. J. Savage,. Agric. For. Meteorol, **148**, (2008)
20. M. F. Li, H. B. Li, P.T. Guo, W. Wu, Energ. Convrs. Manage, **51**, (2010)
21. G. N. Okonkwo, A. O. C. Nwokoge, ESJ, **10**(15), (2014)
22. M. G. Abraha, M. J. Savage, Agric. For. Meteorol, **148**, (2008)
23. V. H. Quej, J. Almorox, M. Ibrakhimov, L.Saito, Energ. Convrs. Manage, **110**, (2016)
24. W. Yao, Z. Li, Y. Wang, F. Yiang, L. Hu, Energy Energ. Convrs. Manage, **84**, (2014)
25. E. A Sahah, F. A. Uba, journal STR, **2**(8), (2013)

 https://doi.org/10.1051/e3sconf/20198001003

Review and resource assessment, solar energy in different region in Ecuador

Juan Lata García[1,2], *Francisco* Jurado[2], and *Víctor* Larco[1]

[1]Department of Electrical Engineering, Universidad Politécnica Salesiana - GIPI, Ecuador
[2]Department of Electrical Engineering, EPS Linares, University of Jaen, Spain

Abstract. Environmental pollution caused by the generation of electricity through fossil fuels leads several countries to adopt strategies for the exploitation of renewable energy sources. In this work, the current energy situation of Ecuador and the incorporation of photovoltaic generators in the national system is reviewed. The document is completed with the evaluation of the energy potential for an average load in the four regions of Ecuador. The selection of the locations of the system under study is chosen through the solar atlas and the meteorological data for a year. The technical dimensioning and economic analysis is achieved through the Homer software. The results show that the insular region reaches the lowest value of levelized cost of energy (COE) $0.529, an autonomy of 22.7 hours and unmet electric load 0.06% per year.

1 Introduction

During the global financial crisis, the substitution of fossil fuels by renewable energy for the production of electricity was promoted [1]. Nowadays, the increase in energy consumption, climate change, emissions derived from fossil fuels and energy independence contribute to the introduction of renewable energies [2]. The use of renewable energy sources has been the policy of several countries in the world to achieve sustainable development [3].

The solar energy is a mature technology and proven in several applications, the energy potential is analyzed by means of climatic and geographical factors for use in urban areas [4]. Latin America is one of the most dynamic markets in terms of renewable energy, the International Renewable Energy Agency (IRENA) indicates that between 2010 and 2015 more than 80,000 million dollars were destined in non-conventional renewable energies, not including large hydroelectric power plants. Brazil, Chile, Mexico, Peru and others are taking advantage of their renewable resources in order to increase their energy and economic efficiency [5].

Ecuador presents the Electrification Master Plan 2013-2022, aligned with the Constitution of the country and the objectives of the National Plan for Good Living, which constitutes a planning tool determining the investments to guarantee the normal supply of electricity, increase of the reliability and security of the electrical system through the development of projects for the generation and expansion of the Ecuadorian electricity sector [6]. In reference [7], the researchers use spatial tools such as Geographic Information Systems (GIS) with the objective of identifying the potential of energy sources in Ecuador for the possible location of solar and wind generators. The results share the areas with greater potential for the implementation of these technologies, the information is classified in spatial layers according to the technology of the renewable resource and location.

Capturing the greatest amount of solar radiation is a complex work due to the constant change in meteorological variables. Several investigations have proposed methodologies to solve the problem, in the reference [8], the authors calculated for a year the azimuth and altitude angles solar. With the information obtained, the sun was tracked on both axes by means of a programmable logic control (PLC), the results showed that with the implemented system of two axes, more energy is collected compared to a fixed system (42.6%). In [9] several configurations of hybrid systems, control strategies and the techno-economic study are analyzed. The authors compare the results using the Homer software and the PSO algorithm. The results show that the feasible solution is based on the COE, renewable fraction, maximum renewable penetration, level cost.

This work studies in section I the number of photovoltaic projects in Ecuador and the sites with the highest radiation in each of the regions of Ecuador, in section II a photovoltaic system is simulated for an average load for the selected places in the section previous, in section III the results obtained are analyzed having as a comparison parameter the cost of energy. Finally, the conclusions are exposed.

2. Background

The multi-year energy balance published by the ARCONEL (Electricity Regulation and Control Agency) shows that Ecuador generated and imported 15,749.65 GWh in 2007, while for 2016 it increased 67.7% to

23,254.62 GWh. The consumption of energy for public service reported in December 2017 is 20,204.22 GWh, which corresponds to 88.66%, the remaining percentage is lost in the distribution system. Table 1 shows the nominal power of the generators installed in Ecuador.

Table 1. Nominal Power in Electric Generation

	Generator	MW	%
Renewable Energy	Hydraulics	4,515.96	56.19%
	Wind	21.15	0.26%
	Photovoltaic	**26.48**	**0.33%**
	Biomass	144.3	1.80%
	Biogas	7.26	0.09%
Total renewable energy		4,715.15	58.67%
Non - Renewable	Thermal	1,937.48	24.11%
	Thermal turbogas	921.85	11.47%
	Thermal turbovapor	461.87	5.75%
Total non-renewable energy		3,321.19	41.33%
Total nominal power		8,036.34	100%

Solar energy had no participation in the Ecuadorian market until 2015 with 0.01 MW. In 2011, the extinct Conelec (National Electricity Council) issues the regulation 004/11 called "Treatment for Energy produced with Non-Conventional Renewable Energy Resources" with the objective of establishing the requirements, prices, their period of validity, and of dispatch for the electric power delivered to the National Interconnected System and systems isolated by the generators that use non-conventional renewable sources, up to an installed capacity of 50 MW [10], The renewable energies included in the regulation are: biomass, wind, photovoltaic, biogas, geothermal and hydroelectric power plants. With which encourages private investment in renewable energy.

The price to be paid in photovoltaic power plants for the energy measured at the point of common coupling (PCC) is 40.03 cUSD/kWh in the continental territory and 44.03 cUSD / kWh in the insular territory of Galapagos, while for wind power plants is 9.13 cUSD/kWh and 10.04 cUSD/kWh respectively. The utilities that subscribed to this contract until December 31, 2015 have a validity of 15 years from the subscription of the qualifying title. Also, they have preferential dispatch in a mandatory manner so that they deliver to the interconnected national system up to the limit of 6% of the installed capacity.

The technical parameters regarding the quality of the electrical energy produced by the photovoltaic systems will be the same established for the conventional generators monitored by the CENACE (National Electricity Operator).

The preferential price attracted the generation companies, where 17 projects were submitted for 284 MW filling the Conelec incentive quota, the expected investment was USD 700 million up to 2015 [11], however the expectations were not met reaching the 26.47 MW of installed power in 2017, distributed in 29 solar plants. Table 2 shows the photovoltaic power plants installed in provinces of Ecuador with their respective power.

Table 2. Nominal power of solar plants

Provinces	Number of plants		Nominal Power (MW)	
	2007	2016	2007	2016
Cotopaxi	0	2	0	2
El Oro	0	6	0	5.99
Galápagos	1	7	0.02	1.64
Guayas	0	4	0	3.98
Imbabura	0	0	0	4
Loja	0	6	0	5.99
Manabí	0	2	0	1.50
Morona Santiago	0	1	0	0.37
Pichincha	0	1	0	1
Total	1	29	0.02	26.47

2.1 Solar radiation in Ecuador

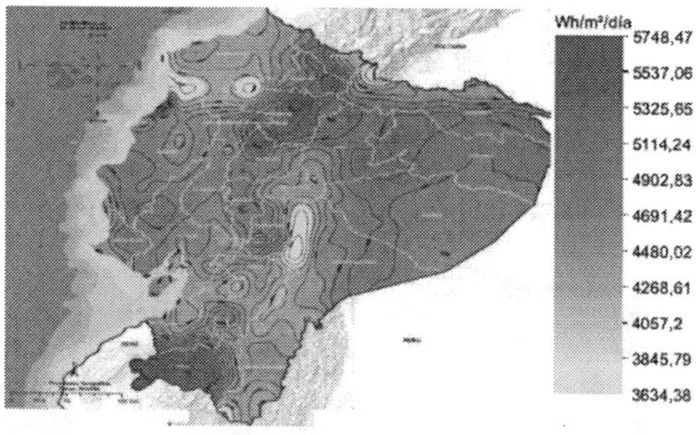

Figure 1. Average global radiation

Solar radiation is radiant energy emitted by the sun, particularly electromagnetic energy, the wavelength and frequency of electromagnetic waves are important to determine their energy. Figure 1 shows the average global radiation map obtained by the solar energy incident on a surface per unit area. For the study of a locality it is necessary to compare the information with meteorological databases (software) and to take field data through meteorological stations. Due to its location and because it is a country with very varied topographic characteristics, Ecuador has a high potential for renewable energy applications. The maximum value of radiation in Ecuador is 5,748 Wh/m²/day, the minimum value is 3,634 Wh/m²/day while the average value is 4,574 Wh/m²/day. The solar atlas was presented by the CONELEC and prepared by the Corporation for Energy Research (CIE) with the aim of promoting the use of solar energy. For the development of the solar atlas the information provided by the National Renewable Energy Laboratory (NREL) of the United States was used, the data have been compared with information from weather stations with an error among the data of 10% [12].

Ecuador has 4 regions (coast, sierra, east and insular region), in each of the regions the highest radiation points are chosen according to figure 1, for the coastal region the province of Manabí is chosen, the coordinates are 0°7.6'S , 80°18.4'W the average annual solar radiation is 4.79 kWh/m²/day. For the sierra region, the province of Loja is chosen with the following coordinates 4°19.8'S, 80°12.7'W the average annual solar radiation is 5.6 kWh/m²/day. For the Ecuadorian east, the place of greatest radiation is in the province of Sucumbíos with the coordinates 0° 27.3'S, 75° 44.'9W with an average annual solar radiation of 4.35 kWh/m²/day. Finally, the last location is for the island region on San Cristobal Island, the coordinates are 0° 53.0'S, 89°35.8'W and the average annual solar radiation is 6.1 kWh/m²/day [13].

3 Proposed Systems

For the simulation of the load, the average per capita consumption 1,143 kWh/inhab was used [14], for a family consisting of 5 members reaching a consumption of 476 kWh/month, the residential load characteristic curve is shown in the figure 2 [15].

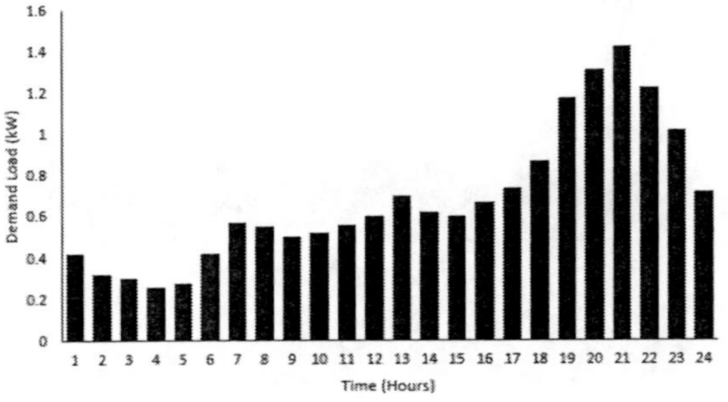

Figure 2. Average global radiation

The load is characterized by a maximum consumption of 1.4 kW, average of 0.66 kW and a minimum consumption of 0.25 kW. The photovoltaic system to be simulated is isolated, as a storage system has a battery bank.

3.1 Photovoltaic Subsystem

A photovoltaic generator consists of several photovoltaic cells connected in series and parallel. The prediction of solar radiation is essential to obtain the generation of the photovoltaic system and know the feasibility of the proposed system. The output power of the photovoltaic generator depends on the solar irradiation, the ambient temperature and is calculated using equation 1.

$$P_{PV} = f_{PV} * Y_{PV} * \frac{I_T}{I_S} \quad (1)$$

Where Y_{PV} is the rated capacity of the PV array (kW), I_T is the total incident radiation on the panel surface (kWh/m²), I_S is 1000 W / m², f_{PV} is the reduction factor that refer to the PV array power output to account for reduced output in operating conditions. The module used for the simulation is of commercial manufacture, flat plate type of 145W, the temperature coefficient is -0.460, operating temperature is 45 °C and efficiency of 14.4%.

3.2 Battery Bank

The storage system is composed of a battery bank. When the production of the photovoltaic panels is greater than the load, the excess energy is stored in the batteries, the storage system is in charge state, the energy stored in a time t can be calculated by equation 2, while the discharge process is calculated by equation 3 [16].

$$E_{Batt}(t) = E_{Batt}(t-1) \times (1-\sigma) + \left[E_{pv}(t) \times \eta_{Inv} - \frac{E_{Load}(t)}{\eta_{Inv}}\right] \times \frac{\eta_{bb}}{\eta_{Inv}} \quad (2)$$

When the energy generated by the photovoltaic system is less than the demand or at night, the battery bank delivers the stored energy to satisfy the load, the process is called discharge and is calculated using equation 3.

$$E_{Batt}(t) = E_{Batt}(t-1) \times (1-\sigma) - \left[\frac{E_{Load}(t)}{\eta_{Inv}} - E_{Gen}(t)\right] \quad (3)$$

Where $E_{Batt}(t), E_{Batt}(t-1)$ are the energy storage in the battery (Wh) in a hour (t); σ is hourly self-discharge rate; E_{pv} is the energy generated by the photovoltaic panel; $E_{Load}(t)$ is the hourly energy required by the load; η_{bb} and η_{Inv} are the battery efficiency and efficiency of the inverter [17].

3.3 Inverter

The system under study contains elements in DC, the balance between the flow of energy between the elements of AC and DC is made by an electronic inverter. The inverter used in this work has a capacity of 2 kW, a life time of 15 years and an efficiency of 95%.

4 Results and discussion

The different configurations are simulated and evaluated using the Homer software, the program evaluates hundreds and even thousands of viable configurations to find the optimal sizing of the hybrid system proposed depending on how the problem is configured, the simulation is performed for one year in steps of time from one hour to one minute.

Taken as reference the data of the solar atlas of Ecuador, the place has been chosen with the highest radiation for each of the regions. The coordinates of the site for the coastal region is 0° 7.1'N, 80° 2.3' W located in the province of Manabí. In the sierra region, the province of Loja has the highest radiation values in the region, the coordinates of the site are 4°19.6'S, 80°17.9'W. The province of Orellana in the Oriente region is chosen for the document, the coordinates of the site is 0° 27.3'S, 75°44.9'W while in the insular region the San Cristobal Island has one of the highest values of radiation with 6.1 kWh/ m²/day, the coordinates of the site is 0° 53.0'S, 89° 35.8'W. The proposed system is observed in figure.3.

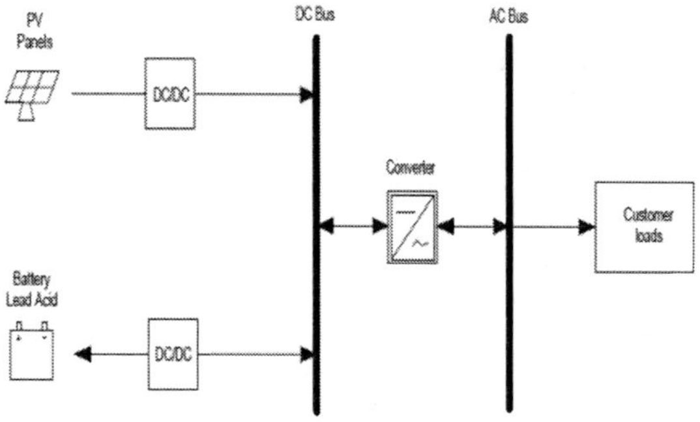

Figure 3. System under study

The results of the simulations are shown in table 3. In the eastern region it has the largest photovoltaic system with 22 kW, 30 batteries of 1 kWh. The cost of energy is $0.659/kWh and the total net present cost is $49,330 which makes it the most expensive system of the evaluated despite the size of the photovoltaic system has unmet electric load of 0.10% (0.608 kWh / yr). Where, the unmet electric load is that load the power system is unable to serve, the battery autonomy is 27.2 h, the expected life is 6.3 yr and the storage wear cost 0.419$/kWh.

The simulated system for the coastal region achieves the following results: a photovoltaic system of 20 kW, 30 batteries of 1 kWh, the cost of energy is of $0.632/kWh and the net present cost is $ 47,322.

The unmet electric load is 0.07 % (4.53 kWh/yr), the battery autonomy is 27.2 hr, the expected life is 6.3 yr and the storage wear cost 0.419 $ / kWh. The results of the battery are the same as those of the previous system when having the same number of batteries.

The simulated system for the Sierra region is a 14 kW photovoltaic system, 28 batteries of 1kWh, the cost of energy is $ 0.571/kWh, the total cost present is $42,748, the unmet electric load is 0.09% (5.69 kWh/yr), the expected life of battery is 5.86 yr and the storage wear cost 0.419 $ / kWh, the battery autonomy is 25.4 hr which is lower than the previous systems evaluated.

The system evaluated with the best result is the one located in the insular region with a 12 kW photovoltaic generator, 25 batteries of 1 kWh, the cost of energy is 0.529, the net present cost is $39,588, the unmet electric load is 0.06% (4.02 kWh/yr), the expected life of battery is 5.26 yr and the storage wear cost 0.419 $/kWh, the battery autonomy is 22.7 hr. The proposed system for the insular region has better economic results, but also less benefits from the bank of batteries because it has fewer units.

Figure 4 shows the total electrical load served for the proposed system in each of the regions, which is the total amount of energy that was used to satisfy the main and deferrable loads during the year. The Costa region has a total power in a year of 5,790.9 kW, in the Insular region it is 5,791.4 kW, in the Oriente region it is 5789.3 kW, while the sierra region is 5,789.7 kW.

The excess electricity in the simulated system for the Galapagos region is 14,774 kW / year which is 68.1%.

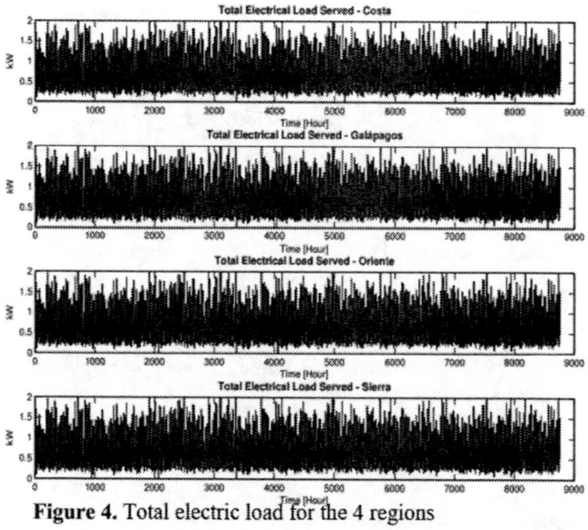

Figure 4. Total electric load for the 4 regions

Table 3. Simulation results of photovoltaic systems

Región	PV (kW)	Battery (kWh)	Converter (kW)	COE ($)	NPC ($)	PV Production (kWh)	Battery Autonomy (hr)	Unmet Electric Load (%)
Oriente	22	30	2	$0.659	$49,330	28,060	27.2	0.10
Costa	20	30	2	$0.632	$47,322	28,314	27.2	0.07
Sierra	14	28	2	$0.571	$42,748	23,161	25.4	0.09
Galápagos	12	25	2	$0.529	$39,588	21,709	22.7	0.06

5 Conclusions

The present work shows the high solar radiation that is in the regions of Ecuador, the highest value of global solar radiation was found in the Galapagos Islands, the annual average value is 6.1 kWh/m²/ day. As a result of the new policies implemented in the electrification master plan, the status changed from the 0.02 MW installed in 2007 of solar plants to 26.47 MW, which contributed 0.33% to the generation of renewable origin in 2016, which represents the 58.67% at the country level. It is still a small percentage, but there have been samples of the implementation of solar generators. The work is completed with the economic technical evaluation for a simulated average load in each of the regions of the country, the results show that in the Galapagos region the lowest COE $ 0.529 and TNPC $39.588 are achieved, where the main generators are of non-renewable origin. This work intends to give an approach to entrepreneurs and the government for the implementation of this type of technology.

References

1. E. K. Stigka, J. A. Paravantis, and G. K. Mihalakakou, "Social acceptance of renewable energy sources: A review of contingent valuation applications," Renew. Sustain. Energy Rev., vol. 32, pp. 100–106, Apr. 2014.
2. S. Fuss, D. J. A. Johansson, J. Szolgayova, and M. Obersteiner, "Impact of climate policy uncertainty on the adoption of electricity generating technologies," Energy Policy, vol. 37, no. 2, pp. 733–743, Feb. 2009.
3. J. Lata-García, C. Reyes-Lopez, and F. Jurado, "Attaining the Energy Sustainability: Analysis of the Ecuadorian Strategy Ku zrównoważoności energetycznej: analiza Strategii Ekwadorskiej," vol. 13, no. 1, pp. 21–29, 2018.
4. M. Ouria and H. Sevinc, "Evaluation of the potential of solar energy utilization in Famagusta, Cyprus," Sustain. Cities Soc., vol. 37, pp. 189–202, Feb. 2018.
5. Banco Mundial, "La energía que necesita la América Latina del futuro." [Online]. Available: http://www.bancomundial.org/es/news/feature/2017/11/23/energias-renovables-america-latina-futuro. [Accessed: 17-Apr-2018].
6. Ministerio de Electricidad y Energía Renovable, "Plan Maestro de Electrificación 2013 - 2022," Quito, 1, 2013.
7. J. Cevallos-Sierra and J. Ramos-Martin, "Spatial assessment of the potential of renewable energy: The case of Ecuador," Renew. Sustain. Energy Rev., vol. 81, pp. 1154–1165, Jan. 2018.
8. C. Sungur, "Multi-axes sun-tracking system with PLC control for photovoltaic panels in Turkey," Renew. Energy, vol. 34, no. 4, pp. 1119–1125, Apr. 2009.
9. Y. Sawle, S. C. Gupta, and A. K. Bohre, "Review of hybrid renewable energy systems with comparative analysis of off-grid hybrid system," Renew. Sustain. Energy Rev., vol. 81, pp. 2217–2235, Jan. 2018.
10. CONELEC, CONELEC 004/11. .
11. Revista Lideres, "17 proyectos para instalar 284 megavatios | Revista Líderes." [Online]. Available: http://www.revistalideres.ec/lideres/17-proyectos-instalar-284-megavatios.html. [Accessed: 19-Apr-2018].
12. Agencia de Regulación y Control de Electricidad, "Estadística del Sector Eléctrico Ecuatoriano 2001," Quito, 2002.
13. S. Podder and M. Z. R. Khan, "Comparison of lead acid and Li-ion battery in solar home system of Bangladesh," in 2016 5th International Conference on Informatics, Electronics and Vision (ICIEV), 2016, pp. 434–438.
14. "Consumo Anual Per Cápita | Agencia de Regulación y Control de Electricidad." [Online]. Available: http://www.regulacionelectrica.gob.ec/estadistica-del-sector-electrico/recaudacion-anual/. [Accessed: 22-Jul-2017].
15. "Plan Maestro de Electrificación 2013- 2022, Volume 2, Pag 42, Residential load curve, Guayaquil electric," 2017. .
16. "Sizing of stand-alone photovoltaic/wind/diesel system with battery and fuel cell storage devices by harmony search algorithm," J. Energy Storage, vol. 2, pp. 30–42, Aug. 2015.
17. "Techno-economic optimization of hybrid photovoltaic/wind/diesel/battery generation in a stand-alone power system," Sol. Energy, vol. 103, pp. 171–182, May 2014.

Identification of factors affecting sustainable development of renewable energy sources in transport

Leontiy Eder[1, 2], *Irina* Filimonova[1, 2], *Vasily* Nemov[1, 2], *Anna* Komarova[1, 2], and *Vladislav* Kozhevin[2]

[1]Trofimuk Institute of Petroleum Geology and Geophysics SB RAS, Russia
[2]Novosibirsk State University, Russia

Abstract. At present motor transport is one of the key components of the formation of a long-term policy of sustainable economic development. Analysis of the structure of energy consumption in motor transport is necessary for better understanding of the factors that affect the increase in the share of renewable energy sources in transport and the effectiveness of specific support measures for the transition to vehicles with minimal or no greenhouse gas emissions. Models of pooled regression, regression with random and fixed effects were built in order to reveal the influence of the chosen factors on the share of renewable energy and electricity in the structure of energy consumption by vehicles. The results of evaluation for 31 European countries in the 2005-2017 showed that the criterion, which allows identifying groups of countries, is the volume of environmental transport taxes per vehicle unit. Countries with higher taxation volumes has such influencing factors as Corruption Perceptions Index, Energy productivity, Total environmental taxes per capita and Employment in high- and medium-high technology manufacturing sectors and knowledge-intensive service sectors. Consumption in countries with lower tax differentiation is influenced by index of research intensity, Corruption Perceptions Index and Shares of environmental and labour taxes.

1 Introduction

Analysis of the present state and structure of energy consumption is a key component in the formation of long-term policies for sustainable development of the economy [1]. Currently, an important aspect is the relationship between motor vehicles and pollution. Reduction of the negative impact on the environment and human health is associated with the improvement of automotive technologies and the transition to alternative fuels in road transport, primarily electricity and renewable sources [2].

In order to reduce the emission of greenhouse gases, European countries have set targets for efficiency of energy consumption, which were achieved, as well as increase the usage of renewable energy sources, thus becoming world leaders in this area.

By 2020, the goal is set for energy consumption to be at least 20% from renewable sources (Directive 2009/28/EC) and at least 27% by 2030 (2030 Energy Strategy). However, in 2016, the consumption of oil products by vehicles in the European Union was 95%. Moreover, according to the European Environment Agency (EEA), the amount of greenhouse gas emissions due to fuel combustion in motor transport increased by 5.4% in the period of 2013-2016.

An analysis of the structure of energy consumption in motor transport is necessary for better understanding of the factors that affect the increase in the share of renewable energy sources in transport and the effectiveness of specific support measures for the transition to vehicles with minimal or no greenhouse gas emissions. In addition, some researchers [3] believe that electrical energy consumption in transport will increase significantly, so its role must be taken into account to adequately provide the population with electricity. For example, in North Europe countries (Denmark, Norway, Iceland, Sweden and Finland), electric vehicle sales increased by 57% in 2017 compared to the previous year [4]. At the same time, the improvement of technologies leads to an increase in the energy efficiency of traditional engines and to a certain reduction in the demand for petroleum products [5, 6]

The object of the study is renewable energy sources in the fuel balance of the motor transport.

The purpose of the study is identification of the factors determining the level and dynamics of consumption of renewable energy resources.

The objectives of the study include: 1) analysis of the main trends in energy consumption on motor transport; 2) identification of factors affecting the level and dynamics of consumption of renewable energy sources; and 3) building of a multifactor model of the influence of the identified factors on the consumption of renewable energy sources using panel data.

Scientific novelty of research consists of identification of main factors affecting the consumption renewable energy sources on road transport, including institutional features of countries, tax incentives and the level of technological development; building of multifactor model based on established and systematized indicators with following verification of the significance of factors.

With the development of infrastructure and the improvement of the operational and economic characteristics of cars using alternative fuels, the attractiveness of renewable energy sources is increasing. The share of biofuels in the structure of energy consumption by motor transport in the world is about 4%. Thus, it represents a large share of the renewable energy in the structure of demand for transport energy. In addition, renewable energy in transport is represented by cars using natural gas and electricity [7]. According to British Petroleum [8], the number of electric vehicles will increase from 1.2 million in 2015 to 100 million in 2035, which is 6% of the world's fleet.

The study of demand for cars using alternative energy sources was carried out by Mabit [9], using such factors as price, annual costs, acceleration time and operating range.

In addition to forecasting demand for cars using alternative fuels, researchers are trying to predict their impact on oil consumption. Oil is the main energy carrier in transport, and the main obstacle to the introduction of renewable fuel. The dependence of investments in alternative energy sources on oil prices, real GDP and interest rates is described with the help of vector regression in the work of Shah, Hiles and Morley [10]. Sadorsky in "Modeling renewable energy consumption for a greener world economy" also uses vector autoregression to analyze the relationship between alternative energy consumption, income, oil consumption and oil prices [11].

As shown in the studies of the International Energy Agency "Nordic EV Outlook 2018" [4], "Global EV Outlook 2018. Towards cross-modal electrification", one of the most important factors affecting the decision to buy a car with low or zero emissions is taxes on the car purchase, such as differentiated VAT rates, vehicle registration tax depending on the amount of CO_2 emissions. Thus, a differentiated level of taxation depending on the environmental characteristics of the car, make it economically feasible to purchase cars with better environmental performance or alternative power units.

To analyze the structure of energy consumption in motor transport and identify significant factors for countries with different levels of environmental taxes on transport, a grouping of countries was carried out. The division into groups is based on the indicator of the level

2 Data

The model is based on data from 31 European countries. The database was formed from the statistics of Eurostat, World Bank's World Development Indicators and International Energy Agency data. The observation period under study is between 2005 and 2017

The following factors that could influence the consumption of renewable energy sources in motor transport were analyzed: the prices for traditional motor fuel, tax incentives, the degree of development of the institutional environment and technologies. Specific indicators that reflect these factors are: prices for traditional motor fuel - the cost of motor gasoline; tax incentives - total volume of environmental taxes per capita and shares of environmental and labor taxes in total tax revenues; the degree of development of the institutional environment - the index of perception of corruption; the level of technology development - the energy efficiency of the economy, share of employed in high- and medium-high technology manufacturing sectors and knowledge-intensive service sectors and the share of R & D in the GDP structure.

Share of renewable energy and electricity in the structure of energy consumption by vehicles was chosen as a dependent variable.

3 Methodology

Models based on panel data are among approaches for building models of the dependence of variables on various factors. The panel data combines both spatial type data and time series data.

In general, the regression model of panel data has the following form [12, 13]:

$$x_{it} = Z_{it} \propto_{it} + \varepsilon_{it} \quad (1)$$

where i denotes index of economic unit, t – time period, $\propto_{it}$ - the coefficients of the explanatory variables of vector Z_{it} in the time period t for the unit i, ε_{it} – vector of errors.

This model cannot be estimated, but it is possible to isolate specific factors that are related to a particular economic object at a certain point in time, consequently individual (specific) features can be evaluated.

$$x_{it} = Z_{it} \propto_{it} + \gamma_t + f_i + \varepsilon_{it} \quad (2)$$

with Z_{it} denoting n-dimensional vector of independent variables without constant, f_i – individual features that do not depend on time, γ_t– specific features that are not observed in all periods, but for all objects, ε_{it} – independent, identically distributed random variables (both in individuals and in time) with zero mathematical expectation and variance σ_ε^2.

Depending on the assumptions about the nature of f_i, models with fixed (f_i are N constant unknown parameters) and random effects (f_i are random and not correlated with errors) are distinguished.

To choose between model with fixed-effect, pooled regression, and model with random effects a Wald test, Breusch–Pagan test and Hausman test are performed.

Once the relationship model of the dependent variable and factors has been obtained, it is necessary to estimate its accuracy with the help of the determination coefficient, showing how well the obtained regression describes the dependent variable and accounting to be at least 50%.

Further, it is necessary to check statistical significance of the obtained regression with the help of F-statistic on 5% significance level. The presence of multicollinearity can be verified using the coefficients of pair correlation. If the factor correlation coefficient is large, then there is a connection between these factors.

4 Empirical testing

Two models were developed on the bases of different dependent variables determining the share of RES in the energy consumption structure in transport sector. The division of the observed countries into two groups was established in the course of the analysis and confirmed by regression calculations with dummy variables. Group one consists of countries with environmental taxes by economic activity per capita greater than 200 euro and second group – less than 200 euro.

All the developed models show robust results with significant F-statistic, presenting significance of the best in-group models – for (I) pooled regression F-statistic equals 30.94, for (II) pooled regression F-statistic equals 14.20.

Based on the data for the European countries and (I) model, three models were built: a pooled regression, model with fixed effects, and model with random effects (Table 1). The model of cross-sectional regression is based on the assumption that all the studied countries are identical, thus, a normal multifactorial regression was built.

Table 1. Assessment of the model for the economies of 12 European countries from group one (significance level 0.05)

Variable	Pooling	Random	Fixed
(Intercept)	-8.44***	-7.50***	
Logarithm (EMPL_INTENS_SECT)	0.71	0.77*	-1.44
Logarithm (ENERGY_PRODUCTIVITY)	0.87***	1.34***	1.93***
Logarithm (CPI)	1.83**	2.26**	0.81
Logarithm (ENV_TAX_PER_CAP)	-0.44	-1.04*	-1.58**
R^2	0.39	0.69	0.34
Adj. R^2	0.34	0.67	0.12
F-statistic	8.69	30.94	5.71

Comparison of the determination coefficient of the models shows total dominance of random effects regression model (R-Squared = 0.69), while fixed effects (R-Squared = 0.34) and polling (R-Squared = 0.39) show lower efficiency, but in general it is not efficient approach. Then the most suitable model was chosen by Wald, Breusch-Pagan and Hausman tests and F-statistic – random effects model (Table 2).

According to the F- statistic, the model is significant (p-value is less than 1.68e-13 for random effects regression). All observed factors are significant at 5% level.

Table 2. Tests of models (significance level 0.05)

Test	Df	Chisq	p-value	Best model
Wald	1	47.34	2.2e-16	Fixed
Breusch-Pagan	1	43.59	4.0e-11	Random
Hausman	4	8.71	0.07	RE and FE

Based on the data for the European countries and (II) model, three models were built: a pooled regression, model with fixed effects, and model with random effects (Table 3).

Table 3. Assessment of the model for the economies of 12 European countries from group two (significance level 0.05)

Variable	Pooling	Random	Fixed
(Intercept)	5.11***	2.74	
Logarithm (RD_GDP)	0.24**	0.24	0.40*
Logarithm (CPI)	-0.51	0.09	0.97*
Logarithm (ENV_TAX_PER_CAP)	-0.85***	-0.87**	-1.55***
R^2	0.41	0.29	0.44
Adj. R^2	0.38	0.25	0.27
F-statistic	14.20	8.11	12.78

Comparison of the determination coefficient of the pooled regression model (R-Squared = 0.41) and fixed effects (R-Squared = 0.44) show that the estimates obtained are much efficient than the estimates of the random effects model (R-Squared = 0.29). Then the most suitable model was chosen by Wald, Breusch-Pagan and Hausman tests – model with fixed effects (Table 4).

According to the F- statistic, the model is significant (p-value is less than 3.98e-07). All observed factors are significant at 5% level.

Table 4. Tests of models (significance level 0.05)

Test	Df	Chisq	p-value	Best model
Wald	1	14.28	1.6e-04	Fixed
Breusch-Pagan	1	44.51	2.53e-11	Random
Hausman	3	1.31	0.72	RE and FE

5 Discussion

The main energy source for motor transport is oil products, which share in total consumption is 94%. At the same time, consumption of alternative fuels such as biofuels, electricity and natural gas is growing. The consumption of renewable energy sources (RES) in motor transport is influenced by various groups of factors: economic, cost, and infrastructure. However, the high cost of cars with zero or low CO_2 emissions is the main limiting factor for increasing the share of renewable energy sources in motor transport.

The analysis showed that the studied countries have a different elasticity of the share of RES in motor transport by the chosen factors. Thus, in the first group of countries there is a high sensitivity of the share of RES to the change in institutional and technological factors, while in the second group of countries the factors under study weakly stimulate the use of RES. The study showed that the criterion, which allows identifying these groups of countries, is the volume of environmental transport taxes per vehicle unit. Ecological taxation in transport includes such payments as Registration or use of motor vehicles, recurrent (e.g. yearly taxes), Road use (e.g. motorway taxes), Congestion charges and city tolls, and others. The average value of environmental taxes per vehicle unit in the surveyed countries ranges from 17 euros (Estonia) to 1700 euros per year (Denmark).

The first group included countries with an average level of environmental taxes of more than 200 euros per car. These are, first of all, the countries of Northern Europe (Denmark, Norway, etc.) and Central Europe (Germany, Netherlands, Belgium, etc.), as well as Great Britain, Croatia and Cyprus. The second group of countries with an average level of environmental taxes of less than 200 euros per car is primarily the countries of Eastern Europe (Czech Republic, Bulgaria, Hungary, Slovakia, etc.), as well as some countries in Southern and Western Europe (Portugal, Spain, Italy) .

Significant factors for the first group of countries were: Corruption Perceptions Index, Energy productivity, Total environmental taxes per capita and Employment in high- and medium-high technology manufacturing sectors and knowledge-intensive service sectors. For the second group of countries, the index of research intensity, Corruption Perceptions Index and Shares of environmental and labor taxes were the significant factors.

Differentiated tax rates (depending on the environmental characteristics of the car) allow for overcoming of the barrier of high initial cost of cars with low or zero greenhouse gas emissions. This leads to a greater variability in the demand for RES in response to a change in various factors.

In countries with a high level of environmental charges, the most significant factor is the degree of development of institutions. Hence, high ecological taxes on motor transport do not always lead to the expected impact on the growth of consumption of renewable energy sources. However, there is a significant increase in the share of RES and reduction of greenhouse gas emissions by vehicles in countries with high quality institutions, the introduction of a taxation system depending on the environmental characteristics of cars, coupled with incentives such as free parking, access to lanes dedicated for buses, exemption from payment for the use of regional toll roads, etc.

The second significant factor for the first group of countries is energy productivity. This indicator characterizes the policy of the state and the effectiveness of measures taken in the matters of energy conservation and ecology. The high efficiency of measures to increase energy efficiency promotes the introduction of advanced technologies in all sectors of the economy and, as a result, the growth of the use of safer energy sources. The third significant factor is total environmental taxes per capita. Higher tax burden per capita on all sectors of the economy in the field of ecology stimulates the use of technologies and energy carriers with the least impact on the environment.

For all groups of countries, the level of technological development of the country is a significant factor represented by the share of renewable energy sources: employment in high- and medium-high technology manufacturing sectors and knowledge-intensive service sectors and the indicator of the science intensity of the economy.

The specific significant factor for the second group of countries is Shares of environmental and labor taxes in total tax revenues. At the same time, an increase in the share of environmental taxes leads to a decrease in the use of renewable energy sources. Thus, in countries where there are no transport taxes that remove the barrier of high initial cost of environmentally safe vehicles, an increase in other environmental taxes results in a reduction of the well-being of the population and increase in the usage of traditional, cheaper types of engines that burn petroleum products.

6 Conclusion

One of the most important factors affecting the use of renewable energy sources for motor transport is the volume of transport environmental taxes per vehicle unit, which allow removing the barrier of high initial cost of high-tech vehicles. In countries with high levels of environmental taxes and differential tax rates, factors such as employment in high- and medium-high technology manufacturing sectors and knowledge-intensive service sectors, the total volume of environmental taxes in the economy per capita have a significant influence on the decision-making in favour of RES in transport. In countries without substantial transport taxes, the specific significant factor for the second group of countries is shares of environmental and labour taxes in total tax revenues. Thus, an increase in other environmental taxes leads to a decrease in the well-being of the population and the use of traditional, cheaper types of engines that burn petroleum products.

The research was supported by the RFBR grant 17-0600537 and grant of the President of Russian Federation MD-6476.2018.6.

References

1. BP Energy Outlook. BP, 2017.
2. I.S. Imran Hussain, C. Hiles, B. Morley, How do oil prices, macroeconomic factors and policies affect the market for renewable energy. Applied Energy, vol. 215(C), 2018.
3. Nordic EV outlook 2018. International Energy Agency, 2018.
4. P. Sadorsky. Modeling Renewable Energy Consumption for a Greener Global Economy. Planet Earth, 2011.
5. S. L. Mabit. Danish preferences concerning alternative-fuel vehicles. Procedings ETC, 2009.
6. Y. He, Q. Zhang, Y. Pang, The development pattern design of Chinese electric vehicles based on the analysis of the critical price of the life cycle cost. Energy policy, vol. 109(C), 2017.
7. L. Eder, I. Provornaya, Analysis of energy intensity trend as a tool for long-term forecasting of energy consumption. Energy Efficiency, 2018.
8. L. Eder, I. Filimonova, V. Nemov, I. Provornaya, Forecasting sustainable development of transport sectors of Russia and EU: Energy consumption and efficiency. International Journal of Energy Economics and Policy, 8(2), 2018.

9. L. Eder, V. Nemov, Forecast of energy consumption of vehicles. Studies on Russian Economic Development, 28(4), 2017.
10. L. Eder, I. Filimonova, V. Nemov, I. Provornaya, Forecasting of energy and petroleum consumption by motor transport in the regions of the Russian Federation. Economy of Region, 13(3), 2017.
11. A. Kontorovich, M. Epov, L. Eder, Long-term and medium-term scenarios and factors in world energy perspectives for the 21st century. Russian Geology and Geophysics, 57(12), 2014.
12. E. Kolomak, Panel data: methods of econometric analysis: textbook. Novosibirsk: Editorial and publishing center of NSU, 2011.
13. T. Ratnikova, K. Furmanov, Analysis of panel data and data on the duration of States: studies manual. Moscow: Publishing house of the Higher School of Economics, 2014.

Use of closured open pit and underground coal mines for energy generation: Application to the Asturias Central Coal Basin (Spain)

Javier Menéndez[1] and *Jorge* Loredo[2]

[1]HUNASER ENERGY, 33005 Oviedo, Spain
[2]UNIVERSITY OF OVIEDO, Mining Exploitation Department, 33004 Oviedo, Spain

Abstract. The idea to reducing the dependency of nuclear and fossil energy sources has risen in Spain, then, the use of closured coal mines for energy generation is thus in the political interest, and by 2020 it is intended that the country produces 20% of its energy from renewable sources, therefore replacing the energy based on fossil fuels. New studies on the framework of a circular economy have led to found new possibilities for closured coal mines in historical mine districts and mine water can be considered as a potential resource, converting mining exploitations into storage and production facilities of renewable energy and generating new economic activities in the mining regions. Abandoned and flooded mines constitute artificial karst type aquifers, and these created underground reservoirs can be economically managed to supply geothermal energy to villages around the shafts. This potential application of mine water, profitable in both economic and environmental terms, could contribute to improve economic and social conditions of traditional mining areas in gradual decline. On the other hand, the restored lands can be used for the production of biofuels through repopulation with fast-growing energy crops and high planting densities. This paper analyses the uses of mine water for energy generation both as geothermal resource and through pumped storage between an underground and a surface reservoir, furthermore the supply of productive restoration for biomass production in closured coal open pits is considered, in relation to their potential application to the closured coal mines from the Asturias Central Coal basin in Northern Spain.

1 Introduction

The contribution of renewable energies to the world's total energy demand has increased particularly during the last two decades, and they will continue gaining market share. The idea to reducing the dependency of nuclear and fossil energy sources has risen in Spain, then, the promotion of renewables is thus in the political interest, and by 2020 it is intended that the country produces 20% of its energy from renewable sources, therefore replacing the energy based on fossil fuels. This paper analyses the potential use of closured coal mines for energy generation.

Even though it is regarded as a crucial economic activity worldwide, mining has a significant negative impact on environment. Due to its nature, especially opencast mining inevitably leads to serious degradation on ecological and aesthetic values of the landscape. When the extraction of reserve is over, the altered landscape has to be reclaimed in order to relieve the damaging effects of opencast mining and restore the landscape and its immediate surroundings. Mining operations alter the natural flow of groundwater, and the mining voids, together with the spaces created due to the fracturing induced by mining, can be filled up with water, once the mines are closed and inundated, materializing then a new aquifer or "mining reservoir". Flooded mines constitute artificial karst type aquifers.

New studies on the framework of a circular economy have led to found new possibilities for mine water which can be considered as a potential resource, converting mining exploitations into storage and production facilities of renewable energy and generating new economic activities in the mining regions. Then, mine water, either coming from the pumps or from the stage rebound after flooding during the closure period, could be used with economic and social benefits for energy generation.

2 The Asturian Central Coal Basin (ACCB) in northern Spain

The Asturian Central Coal Basin (ACCB), which has been exploited from the end of the eighteen century and the lst mine in operation wil be closured for the end of the present year. This coal mining district constitutes the main Spanish coal basin and it is characterized from a hydrogeological point of view by the presence of predominantly low porosity and permeability materials. Groundwater flow occurs mainly through mining voids, open fractures, and zones of decompression associated with coal mining.

The Asturian Central Coal Basin (ACCB) is located in northern Spain. With an area of about 1400 km², the ACCB is the largest carboniferous outcrop of the peninsula and the main Spanish coal mining district. This coal basin has been an exploited for more than 200 years, through open pit and underground mining, with indoor mining predominating in the last decades.

This potential application of mine water, profitable in both economic and environmental terms, could contribute to improve economic and social conditions of traditional mining areas in gradual decline.

2.1 Potential for energy generation in closured underground mines

When mining ends, one option is to abandon pumping, gradually flooding both the mine voids and the open pores; this 'groundwater rebound' [1] continues until potentiometric equilibrium is reached [2]. If the lithological units connected to any part of the flooded system intersect the surface at a level lower than that of the mouth of the mine shaft, water upwelling (a spring) will occur. If pumping is resumed such that its discharge is equal to the recharge, the water level can be stabilized throughout the ductwork of the new aquifer. The stored volume is then equivalent to that of the filled voids and an underground reservoir has been created. These created underground reservoirs can be economically managed to supply geothermal energy (mainly by means of heat pumps) and hydroelectricity by Pumped Storage Hydropower (PSH) projects with a Francis turbine-pump.

2.1.1 Mine water as a geothermal resource

One of the most important uses of mine water is the production of thermal energy for air conditioning of buildings. The temperature of water from the coal mining reservoirs in Europe is usually above 14ºC, so modern heat pumps using it as cold source are competitive, considering current prices of electricity and fuel [3]. Producing energy by means of heat pumps using mine water, compared to conventional systems using natural gas, in several applications, means economic savings above 70% and reductions of CO_2 emissions between 20-40%.

Heat pump technology has been around since 1852 when Lord Kelvin first developed the concept. In the 1940s, Robert Webber modified the concept using the ground as the source of heat. These ground source or geothermal heat pumps systems gained popularity in the 1960s and 1970s. Figure 1 shows a diagram of a heat pump system. In heating mode, the refrigerant in a cold liquid form gains heat from the outside source (air or ground) in a heat exchanger (evaporator), where it is converted into a cold vapor. In order for the heat pump to function properly, the liquid form of the refrigerant must be at a lower temperature than the outside heat source and the temperature of the heat source must be high enough to cause the refrigerant to evaporate (boil). After the liquid absorbs heat and is converted to a vapor, it is then compressed (requiring an input of electrical energy), converting it to a hot vapor. The hot vapor is sent to another heat exchanger (condenser). Here the hot vapour gives up the heat that was gained from the source in the evaporator and in the process is condensed to a hot liquid (The heat given up is what is used to heat the interior space). The hot liquid goes through an expansion valve where the drop in pressure converts it to a cold liquid and the process is repeated. In heating mode, the evaporator is placed in contact with the heat source.

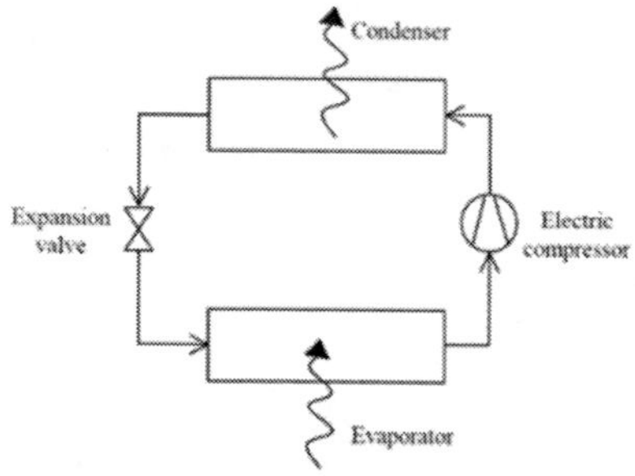

Figure 1. Schematic of head pump system

The heat pumps operate on electricity and the amount of heat generated divided by the amount of energy needed to operate the heat pump is known as the coefficient of performance (COP). The smaller the difference in temperature between the heat source and the heat sink (required temperature of space to be heated), the greater the COP of the heat pump system. For this reason, geothermal heat pumps have a greater efficiency than air-source heat pumps; the source/sink temperature (the earth) is closer to the desired room conditions than outdoor air temperatures.

In ACCB there is an important geothermal project in order to take advantage of the mine water supplying heating and cooling to the surrounding population and industrial districts by use of heat pump technologies.

The mine water stored in the mining reservoir has a considerable geothermal potential, given its temperature (23 °C). It is non-acidic water, but is mineralized and hard, with quite high Fe and Mn content. For heat pump use, the heavy load of dissolved matter that coal mine waters commonly contain can cause problems by precipitating during heating and cooling, so effective (and expensive) water softener and descaling treatment may be necessary. The pumping is necessary to maintain the security level to avoid the flooding of the buildings that are located in the surroundings. Fig. 2 shows the temperature profile of the water stored in the closed coal mines.

 https://doi.org/10.1051/e3sconf/20198001005

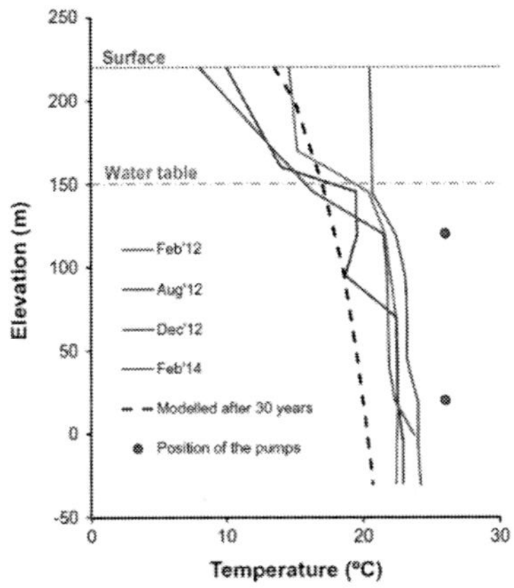

Figure 2. Mining reservoir temperature

Figure 3 shows the thermal energy produced from mine water for the air conditioning of a public building in the ACCB. It is an installation with an installed capacity of 3500 kW to supply cold and heat by means of heat pumps. The coefficient of performance (COP) is also shown, based on the electrical energy consumption of the heat pumps.

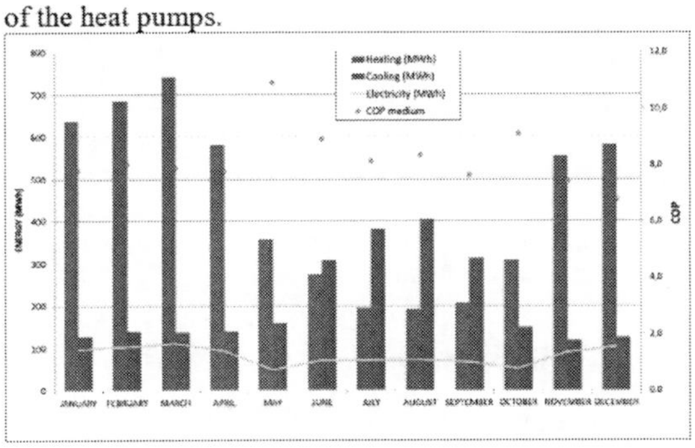

Figure 3. Thermal energy and coefficient of performance

2.1.2 Pumped Storage Hydroelectricity

The tunnel network of a mining facility has an unusual geometry for a storage system. Nevertheless, such storage volumes, combined with the depths at which some of these facilities are located and the high flows of mine water, can be sufficiently appealing to establish a Pumped Storage Hydropower (PSH) project with a Francis turbine-pump. The most relevant technical aspect is related to the storage structure for the lower reservoir. Fig. 4 shows the diagram of the hydroelectric power station. The upper reservoir is located outside and the lower reservoir is located inside the coal mine. The penstock would be built in the vertical shaft currently in existence.

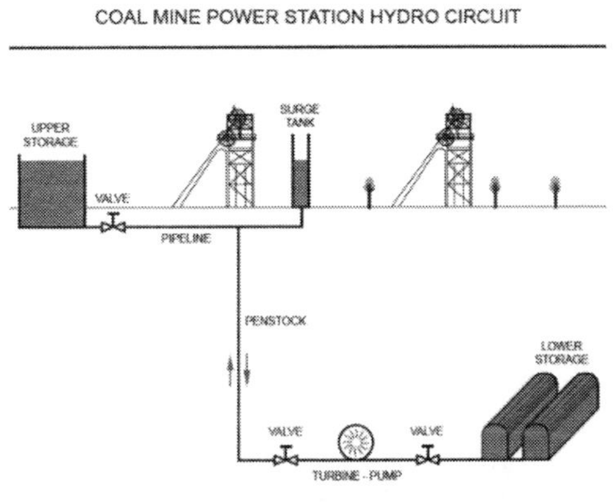

Figure 4. Coal mine hydro power station hydrocircuit

In a typical hydropower pumped storage project, the two reservoirs are located on the surface level. In contrast to a conventional PSHP plant, the upper reservoir of an Underground PSH power plant is the smaller problem, as it can basically be established on the surface. If an abandoned coal mine is envisaged, the (potentially large) area of the former mine may be available for use In the generally densely populated Asturian Central Coal Basin, at least small- and medium-sized storage reservoirs on the surface, may often be done without too much conflict with settlement areas. In the surroundings of the shafts there exist buildings that are protected as industrial heritage, which cannot be demolished.

Pumped hydroelectric energy storage is a large, mature, and commercial utility-scale technology currently used at many locations in the world. Pumped hydro employs off-peak electricity to pump water from a reservoir up to another reservoir at a higher elevation. When electricity is needed, water is released from the upper reservoir through a hydroelectric turbine into the lower reservoir to generate electricity.

Because most low-carbon electricity resources cannot flexibly adjust their output to match fluctuating power demands, there is an increasing need for bulk electricity storage due to increasing adoption of intermittent renewable energy. This technology can be the backbone of a reliable renewable electricity system.

The first idea of exploiting a disused mine as an underground reservoir dated from 1960 [4] and it was developed by several studies and technical reports but not accompanied by functioning pilot projects.

Mount Hope project, located in northern New Jersey was initially proposed in 1975. It intended to use the facilities of an abandoned iron mine as a lower reservoir but it was never developed [5].

The feasibility of using some of the current coal mining facilities in the Ruhr region as lower reservoir for a pumped storage project has been currently analyzed by a group of five partners in Germany (University Duisburg Essen, Ruhr University Bochum, Rhine Ruhr Institute for Social Research and Political Consultancy RISP, RAG AG and DMT), supported by the European Union.

 https://doi.org/10.1051/e3sconf/20198001005

Madlener and Specht presented an extremely interesting techno-economical analysis of the possible construction of underground PHES in Abandoned Coal Mines in the Ruhr area (Germany) [6]. Also in the Ruhr area, Alvarado et al presented a project of the possible construction of underground pumped storage power plant in Prosper Haniel mine in Bottrop (North-Rhine Wesphalia), using existing coal mine infrastructure [7].

This storage concept presents several advantages in comparison with conventional PHES, as for example the higher possibility of social acceptance and the larger number of potential sites. From a technical point of view, even though the construction of an underground storage reservoir is possible, the main limit is the need of competent rock, especially at reservoir depths.

An interesting unconventional pumped hydro project, proposed in Estonia (Project ENE 1001, 2010), is that of Muuga whose completion is expected in 2020. The peculiarity of this project is that it combines two different unconventional reservoirs: the sea as upper reservoir and underground chambers, resulting from granite excavations, as lower reservoir.

As regards the worldwide situation, with over 150 GW, pumped hydro storage power plants represent around 99% of the world's electrical energy storage capacity. Currently Japan is the worldwide leader but China expands quickly and is expected to surpass Japan in 2018. In the future, as the renewable revolution gains momentum worldwide, hydropower looks to become an even more strategic player. The International Renewable Energy Agency (IRENA) conducted a technology roadmap (Remap) until 2030, and hydro capacity could increase up to 60%, and the pumped hydro capacity could be doubled to 325 GW from the 150 GW installed in 2014. In the case of PSH plants (electrical energy), the conclusion for lower reservoirs is that the use of natural caverns is not possible; the artificial extraction of large cavities is technically demanding and financially expensive and thus does not seem to be very reasonable. In certain cases, existing drifts may at least be partly usable, e.g. after additional extension measures. However, for a general concept, considerations have to be based on the fact that the drifts for a rib-shaped storage system in the completion stage, would have to be built totally new. The power of the turbine depends on the flow and the net head. Fig. 5 shows the turbine power for a 6-hour full load cycle time. The power increases as the net head and the volume of water stored in the reservoirs increase.

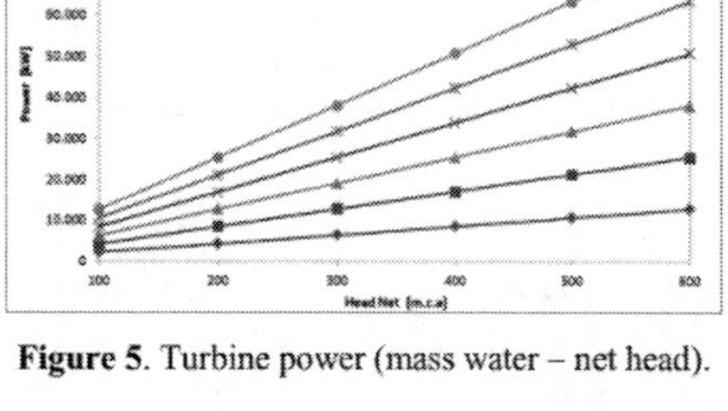

Figure 5. Turbine power (mass water – net head).

2.2 Potential for energy generation in closured open pit mines

During mining, significant amounts of soil and overburden have to be moved in order to access the coal seams beneath. In addition to the alteration in the water network, open-pit mining produces an important impact on the landscape, with visual impacts that must be restored to the end of the extractive activity. The restoration of affected lands is one of the main costs in open-pit mining. The first phase consists of the topographic restitution and the creation of a water drainage network that gives stability to the restored lands, avoiding erosion phenomena as a result of the circulation of runoff water. Later a layer of vegetal earth is added and it is finished with the hydroseeding of the lands. One of the main objectives of the restoration of lands affected by mining activity is the improvement of soil properties. The restoration can be approached from two points of view: environmental or protective restoration, and productive restoration. The environmental restoration is based on the repopulation with autochthonous arboreal species, with a slower growth, and a short-term economic benefit is not sought. On the other hand, through productive restoration, short-term restored land is valued by planting fast-growing energy crops. With this type of crops, much higher growth is achieved and by applying shorter cutting shifts, economic benefits are obtained from the first years using the biofuels obtained for the production of energy. Although the initial investment is higher in the case of productive restoration, the profitability is much higher than in the case of environmental restoration.

2.2.1 *Biomass production*

To reduce carbon dioxide emissions and the dependency on fossil fuels, the developed countries have begiun expanding renewable energy sources to meet energy mix policy goals for 2030 [8]. In 2008 research began on a restored surface of old mining operations in the Asturias Central Coal Basin, with an extension of 7.5 ha that were planted with forest energy crops using different clones of poplar, willow and birch. In total, 3 trials were carried out in different recovery areas. The study area is characterized by an oceanic climate, with average annual temperature of 13°C and an average annual precipitation of 1,115 mm, of which 345 mm falls during the growing season (May–September). Soil formation is at an early stage and the soil structure is still unstable. The steep slopes of the terrain minimize groundwater effects. The physiography of the plots was characterized by a mean slope of 19% and an elevation ranging from 508 to 597 m above sea level.

The first of the plantations was carried out in 2008, the factors to be studied in this first trial were: the type of clone, the density of plantation (two levels: 10,000 and 15,000 plants ha^{-1}) and the treatment (F0=Control F1=300 kg ha^{-1} NPK 6:20:12 and 4 l ha^{-1} glyphosate, F2 = 600 kg ha^{-1} NPK 6:20:12 and 4 l ha^{-1} glyphosate).

The amounts of fertilizer to be used were chosen in view of the results obtained in the soil analysis prior to planting, taking into account the maximum amount of nitrogen allowed to be applied in vulnerable areas, which is 170 Mg ha^{-1}.

3 Conclusions

The exploitation of coal in the Asturian Central Coal Basin in the North of Spain has occupied large areas of land that are currently being restored according to the current legislation.

Closed and inundated coal mines can constitute mining reservoirs which could be regulated and used for energy production, and for theses uses is essential to define the hydrogeological model, the water balance, and the volume of voids of the mining reservoir system previously the design of the project. Considering the pumping rates and the high COP values that can be reached for mine water in the ACCB, their energy use by means of water-water heat pumps for heating and cooling is optimal. The entire coal mining reservoirs in the ACCB involve a potential of energy supply higher than 250 thermal GWh·year^{-1}. Compared to conventional systems, economic savings and a reduction in CO_2 emissions are achieved.

The implementation of an underground PSH project using coal mine facilities is an appealing option for energy storage, particularly in Spain where the underground mining is currently phased out, with an expected closure date at the end of 2018. PSH can provide many services to the power system, as flexibility of operation and speed to vary the power delivered to the grid. This aspect is fundamental to deal with the variations in production due to fortuitous failures in the thermal power plants and of any significant variations in the production of intermittent renewable power generation.

In relation to the restored lands, they can be used for the production of biofuels through repopulation with fast-growing energy crops and high planting densities. Certain poplar clones seem to grow rather well on a degraded mining soil with high pH, even when slightly polluted with heavy metals. Organic fertilization was more efficient than the contribution of mineral fertilizer, also producing a continuous supply of nutrients over longer time and reducing the risk of washing.

In addition to the production of renewable energy and capture of CO_2 emissions, these activities involve the generation of a new economic activity in abandoned land and the creation of jobs in depressed areas due to the closure of mining operations. These applications are encouraging and might be profitably extended to other minor coal mining basins in Asturias and other areas.

References

1. Henton MP. The problem of water table rebound alter mining activity and its effects on ground and surface water quality. In: van Duijvenbooden W, Glasbergen P, Van Lelyveld H (eds) Quality of Groundwater, Proceedings of the International Symposium Noordwijkerhout. Elsevier, The Netherlands, pp 111–116. (1981).
2. Younger PL, Banwart SA, Hedin RS. Mine Water. Hydrology, Pollution, Remediation. Kluwer Academic Publishers, Dordrecht, The Netherlands. (2002).
3. Ochsner K. Geothermal Heat Pumps. A guide to Planning & Installing. Earthscan, London.
4. Harza, R.D. Hydro and pumped storage for peaking, Power Eng., vol. **64** (10), pp. 79–82. (1960).
5. Dames and Moore. An assessment of hydroelectric pumped storage. National hydroelectric power resources study. The U.S. Army Engineer Institute for water resources, pp. A-95-96. (1981).
6. Madlener, R., Specht, J.M. An Exploratory Economic Analysis of Underground Pumped Storage Hydro Power Plants in Abandoned Coal Mines, FCN Working Paper No. 2/2013. (2013).
7. Alvarado, R., Niemann, A., Wortberg, T. Underground pumped-storage hydroelectricity using existing coal mining infrastructure. In: E-proceedings of the 36th IAHR world congress, 28 June–3 July, 2015. The Netherlands: The Hague. (2015)
8. N. Agung, K. Itaoka, A. Chapman, N. Dinh Hoa, N. Yamakawa. Biomass energy in Japan: Current status and future potential. IJSGCE. **Vol 6**, no 2, pp 119-126. (2017).

E3S Web of Conferences **80**, 01006 (2019) https://doi.org/10.1051/e3sconf/20198001006
REEE 2018

The main geothermal reservoirs in the south Algeria and their interest for local economic development

Abdelkader Ait-Ouali[1], *Salima* Ouali[1], *MM* Hadjiat[1] and *Khaled* Imessad[1]

[1]Centre de Développement des Energies Renouvelables, BP. 62 Route de l'Observatoire Bouzareah 16340 Algiers, Algeria

Abstract. The study area is one of the important geothermal provinces in south Algeria. It is characterized by a hot arid climate with intense dryness and very high evaporation rates. The Albian geothermal system is exploited by the wells mainly for domestic and agricultural purposes. The sandstone Continental Intercalary (CI) formation constitutes the reservoir for the Albian aquifer, covering an area of 600,000 km^2. This reservoir is covered by calcareous formations which yield the chemical characteristics of highly mineralized Na-Cl type representing the deep thermal waters and Ca-SO4 type determined the presence of evaporate lithology . For a better understanding of geothermal reservoir characteristic, a multidisciplinary approach was adopted, including hydrogeochemistry and geothermometry. More than fifty samples collected from wells recently in 2017 with a temperature average between 38 and 60° C and conductivities range from 2010 to 3460 µS/cm. Geochemical analysis of those thermal waters exhibits a certain degree of salinity with sodium-chloride type waters domination. The maximal geothermal reservoir temperature estimated using geothermometry is about 120°C.

1 Introduction

Sahara is situated in North Africa between the Atlantic Ocean and the Red Sea, it is the biggest and one of the driest deserted of the planet (Fig. 1). It's up to the big desert located in the north hemisphere, in recent years; several studies address the groundwater and geothermal potential along the south of Algeria]. [1, 2, 3, 4].The natural limits of this vast region of about 900 000 km ^{2} are the following ones:

- In North, the edge of the high plateaus of the Atlas which extends from Morocco to the gulf Gabès gulf (Tunisia). This line corresponds to a geological undulation which brutally stops.
- Towards the south mountains and high plateaus of the Atlas and mark the beginning of the Saharan platform.

In the south, primary formations outcrop bordering the Tademaït and Tinrhert plates.

- In the West, the valley of Saoura which curves between the mounts of Ougarta and Large Erg Occidental and prolongs towards the south by the Messaoud Wadi whose bed is

encumbered sand banks and sebkhas between Adrar and Reggane.

- At the East, the Djebel Dahar reliefs (Tunisia) relayed towards the south by the Algeria-Libyan border. In the summit there is a the multilayer aquifer of Terminal Continental constituted byMio-Pliocene formation and the upper Cretaceous, below, the water table of Continental Intercalary which develops in the Albien and the barremiensablo-clayey lithology. The geothermal systems occur in regions of anomalously high crustal heat flow that may be related to the presence of young igneous bodies or hot rocks located deeper in the crust [2,5]. Frequent data were presented on geothermal waters in South Algeria obtained from fileds works prospection. The South Algerian geothermal fields deal an important setting of high temperature with various chemical proprieties. The distribution of sampling locations is associated with the geology structure cover the study area. The heating origin for South Algerian hydrothermal systems is the high geothermal anomaly related to the average gradient of 4.5°C/100 m [2]. Hydro chemical methods play a key role in geothermal reservoir study and are particularly widely used to determine the deep reservoir proprieties [6, 7]. The current paper focuses on the investigation of chemical properties of the principal geothermal waters located Northern Sahara provinces. Based on recent sampling campaign field work and using adapted measurement tools (WTW multi-parameter tool (2C10-0011FB),...), various data (pH, T°, electrical conductivity (EC),...) were collected and analyzed. This work leads us to provide a reservoir temperature and the corresponding thermal power potential. All results will aid local authorities and private sector investment in various activities (residential, agricultural and tourism) to develop renewable geothermal energy for heat and electricity generation.

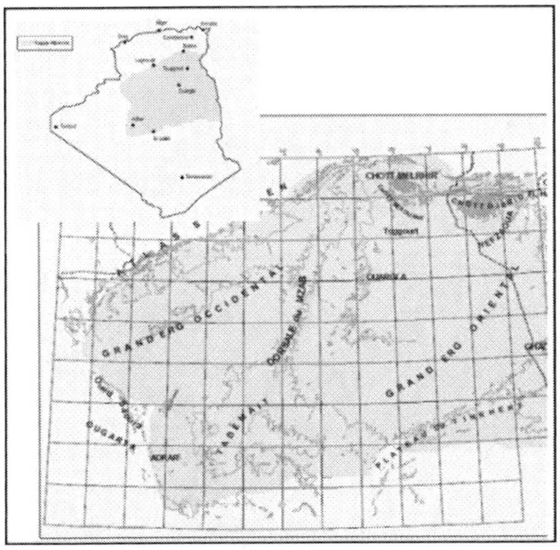

Fig. 1 The huge geothermal reservoir system in the Algerian North Sahara

2 Study area

2.1 Geology aspect

The geology illustrates the outcrops lithology and position in the Sahara (Fig. 2). As well as the typical provision of principal formation the Sahara aquifer system. The Final Complex formations are characterizing by less tectonics deformations. All surfaces which limit the various formations are horizontal [2]. We can subdivide the Algerian Sahara in two under-fields: the Western basin and the Eastern basin, separated by the M' Zab anticlinal from. The Western Sahara basin is the area ranging between Me Zab, the valley Saoura and the atlasique flexure. It's a punt area forms very monotonous, where one observes only one light raising of the Mio-Pliocene layers, so that the whole basin constitutes very open synclinal [1, 5]. More the half Western basin, between Me Zab and Saoura, except for the zone located immediately at the south of the Saharan flexure, we note the absence of intermediate formations between sands of Mio-Pliocene and sands of the Continental Guide. The basin of the Eastern Sahara is definitely better known than its Western counterpart, thanks to oil research [2, 8]. The Eastern Sahara, limited to the west by Me Zab and the south by the plates of Tademait and Tinrhert, is prolonged towards the east in Tunisia and Libya. The Eastern basin is also presented in the form of vast synclinal, open towards north [9, 10].

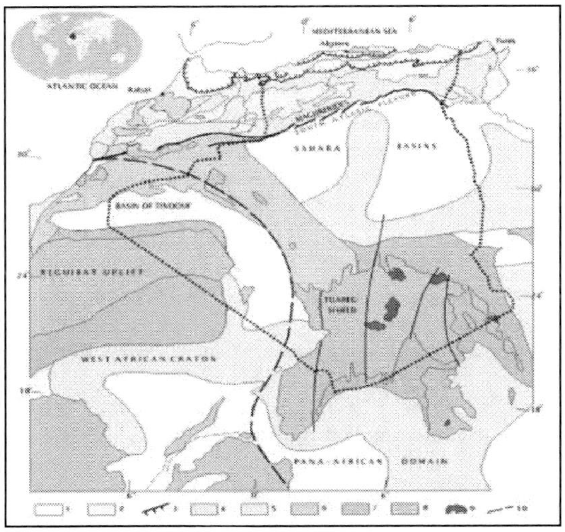

Fig. 2 Major geotectonics units of West Africa modified from Fabre (1976). 1: Tertiary and Quaternary; 2: Alpine molasses; 3: Tertiary thrust sheet; 4: Secondary tabular; 5: Secondary plicative; 6: Primary plicative; 7: Primary tabular; 8: Precambrian and Precorce Cambrian of Sahara; 9: Cenozoic magma; 10: Megafault

2.2 Hydrological contexts

The Algerian Sahara, partly, is occupied by a vast sedimentary basin which is prolonged in Tunisia and Libya. As of the first geological investigations, one recognized that this basin was heterogeneous and contained several very wide fluvio-continental formations, separated by formations considered impermeable. Two principal systems are distinguished: Continental Guide (Ci) and the Final Complex (CT). The Continental Terminal corresponded only to sands and clays of Mio-Pliocene. Under the term of Continental Terminal, with an aim of hydrogeologic simplification from the point of view, all the series going of Cenomanien in Mio-Pliocene. Indeed, the aquifer of the Continental Terminal contained in sands of Mio-Pliocene. The Continental term is used to indicate a hydrogeologic unit and not with its usual direction. The Final Complex includes the most recent formations deposited in the Low-Sahara and limited to the west by the dorsal of Me Zab, in north by the major accident of the Saharian Atlas, in the east by Dahar, the south by a line passing to the north of the axis In Salah-Zarzaitine under Eastern Erg. Sénonien, the Eocene and Mio-Pliocène are, with the Quaternary one, the last formations deposited in the Sahara [11].

3 Methodology

More than fifty sampling sites were investigated with water sampling within North Sahara provinces in 2016. Thermal water samples were collected into 500-ml polyethylene bottles, cleaned with distilled water before collecting the thermal fluid. During the field works campaign, physicochemical parameters such as water temperature, pH, electrical conductivity, Total Dissolved Solids (TDS) are measured, in situ, using WTW multi-parameter tool (2C10-0011FB). During laboratory analysis, Potassium (K+), Calcium (Ca2+), sodium (Na+), Magnesium (Mg2+), Sulfat (SO42-), chloride (Cl-), were determined by ion chromatography (IC, DionexDX-120). The total silica in water was determined by a colorimetric method and alkalinity (HCO3) analyses were made using standard titration technique. Atomic absorption spectrophotometry was used for SiO2 analysis. Chemical analyses of thermal waters were performed at CRD" Centre de Rechreche et development Sonatrach(national oil company).For chemical plotting, the free software "Diagramme"-copyright, provided by Avignon hydrogeology laboratory (http://www.lha.univ-avignon.fr/LHA-Logiciels.htm), was used. Some maps were prepared using the public domain Generic mapping Tool (GMT) software [12]. Several chemical geothermometers were used to estimate the geothermal reservoir temperature [13, 14, 15, 16,17 and 18].

4 Results and discussion

4.1 Geothermal water chemistry

The hydro chemical features of thermal waters are connected to their underground pathway, temperature at depth, the nature of the contact rocks and the residence time. The thermal spring waters are of a variety of chemical types; this variability is due to the lithological composition of the collection sites, which overlie several different complex geological structures. The Piper diagram (Fig. 3) shows that the overall chemical character falls within the following two water types. Most waters belong to the sodium-chloride type Na-Cl. In general, all these waters are sodium-chloride waters as a result of the physical environment or of reactions with deep reservoir lithology (water–rock interaction). Allowing to the Piper diagram, the evaporates layers appear to influence the composition of the circulating shallow waters. A characteristic SO4-type composition was recognized for the study area waters, maybe due to the presence of gypsum and/or anhydrite (CaSO4) in the reservoir. Very high SO4concentrations of some samples advise a carbonate source with anhydrite. A small amount of H2S and CO2 can be released during the bacterial reduction of SO4 and hydrocarbons.

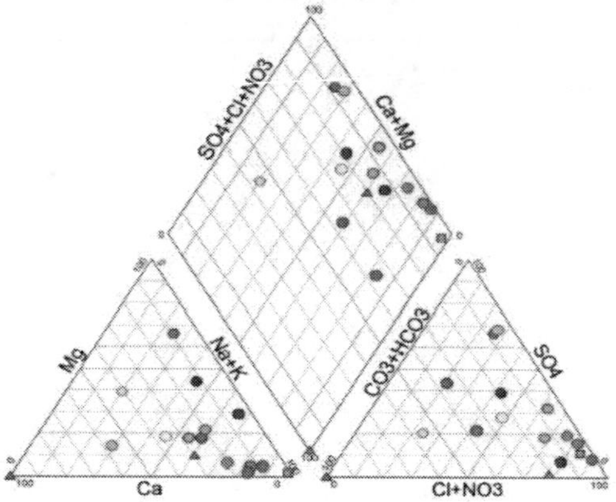

Fig. 3 Piper classification diagram showing the chemistry distribution of the water study area

4.2 Geothermal reservoir temperature

More than forty thermal waters samples selected to assess reservoir temperature by numerous chemical geothermometers. This technique assumes that there is a fluid-mineral equilibrium as well as an equilibration between mineral assemblages at depth in the reservoir and stabilization of dissolved chemical elements with the ascent before mixing. The outcomes of the diverse chemistry methods show that deep reservoir temperatures are quite variable. Some of the results obtained especially for the chalcedony and the Na/K geothermometers are obviously wrong since the calculated temperatures are lower than the measured discharge temperatures. This outcome may be the consequence of the interaction of thermal waters with reservoir rocks during their rising to the surface and the increase of solubility of certain elements with decreasing temperature. The estimated reservoir temperatures range from 60 to 115 °C for the quartz geothermometer, while the cation geothermometer gives the highest value around 140 °C. The wide range of the calculated values reveals that these temperatures cannot be considered reliable. This could be explained by a strong possibility of mixing between fresh and deep waters during upflow from the reservoir to the surface. Also the water composition is greatly affected by the interaction with the evaporitic rocks that are omnipresent in the region. Temperatures calculated by those geothermometers for main thermal sites in the study area given in figure 4. The outcome of the diverse geothermometers show that the deep reservoir temperatures ranged between 65°C and 155°C.

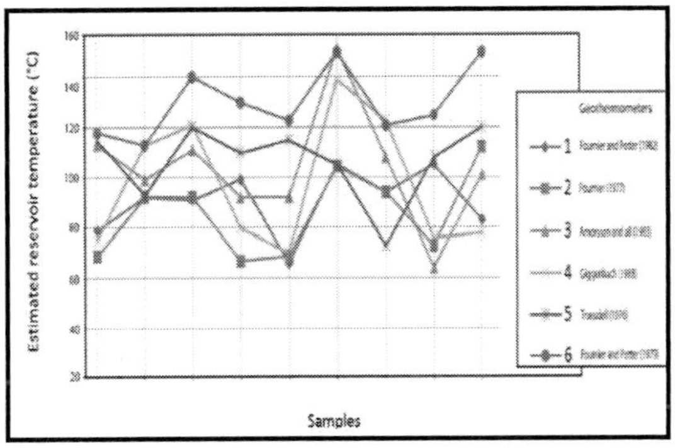

Fig. 4 Estimated reservoir temperature based on different geothermometers

5 Conclusion

The study of the North Sahara geothermal reservoir was carried out in order to investigate the origin and sources of solutes and estimate the deep reservoir temperatures. It examines the hydrochemical composition of more than fifty thermal water samples from springs and wells, with various methods such as geothermomtry analysis, saturation index and Piper diagram. The arrangement of major elements geochemistry has provided a comprehensive understanding of the mineralization processes that reinforce the geochemical evolution of the thermal water inside the study area. The combination of the geological and hydrogeological methods let us conclude a good relation between chemical processes and hydrothermal flow system. The main studied geothermal provinces characterized by an important outlet temperatures ranging from 40 to 62 °C (springs and wells). The chemical analyses data (recent decade field works) show us a deep flow in the Albien sandstone aquifer, influenced by Trias evaporate lithology. The data treatment let us complete that the thermal waters in the study area come from deep geothermal reservoir (1500m to 2000m). Hydrochemistry results suggest that 90 % of water samples are chloride sodic chemical types. The geothermometry applications give a deep flow path and reservoir temperature ranging between 50 °C to 120 °C.

References

1. Saibi, H., 2009. Geothermal resources in Algeria, Renewable and Sustainable Energy Reviews, 13 (9) 2544-2552.
2. Takherist, D., Lesquer, A., 1989. Détermination du flux de chaleur en Algérie. Canadian Journal of Earth Sciences. 26, 615-626.
3. Kedaid, F.Z., 2007. Data base of the geothermal resources of Algeria. Geothermics. 36, 265–275.
4. Fekraoui, A., 1988. Geothermal resources in Algeria and their possible use, Geothermics, 17 (2/3), 515-519.
5. Ouali S., 2016. Etude géothermique du Sud de l'Algérie. Mem. Magister, Université M'Hamed-Bouguerra - Boumerdes; 2006.
6. Giggenbach, W.F., Glover, R.B., 1992.Tectonic regime and major processes governing the chemistry of water and gas discharges from the rotorua geothermal field, New Zealand.Geothermics,21, 121-140.
7. Saibi, H., Ehara, S., 2010. Temperature and chemical changes in the fluids of the Obama geothermal field (SW Japan) in response to field utilization.Geothermics. 39, 228-241.
8. Bouillin, J.P., 1986. Le bassin maghrébin : une ancienne limite entre l'Europe et l'Afrique à l'Ouest des Alpes, Bull. Soc. géol. France. 8 (4) 547–558.
9. Verdeil, P., 1982. Algerian thermalism in its geostructural setting. How hydrogeology has helped in the elucidation of Algeria's deep seated structure. J Hydrol. 56, 107–17.
10. Wildi, W., 1983. La chaîne tello-rifaine (Algérie, Maroc, Tunisie): Structure, stratigraphie et évolution du Trias au Miocène, Rev. Géol. Dyn. Géogr Phys. 24, 201–297.
11. Wessel, P., Smith, H.F., 1998. New, improved version of the generic mapping tools released. EOS Trans. AGU 79, 579
12. Takherist, D., 1986. Etude du flux de chaleur en Algérie. DEA, Univ. of Languedoc, France.
13. Wessel, P., Smith, H.F., 1998. New, improved version of the generic mapping tools released. EOS Trans. AGU 79, 579.
14. Fournier RO., 1977. Chemical Geothermometers and Mixing Models for Geothermal Systems. Geothermics; 5: 41-50.
15. Haouchim, F-Z., 1998.Application of chemicalgeothermometers to thermal springs of the Maghreb, north Africa, Geothermics, 27 (2) 211-233.
16. Nicholson, K., 1993.Geothermal fluids, chemistry and exploration techniques, Book, Springer, 253 pp.
17. Fournier, RO., 1989. Geochemistry and Dynamics of the Yellowstone National park hydrothermal system. Ann. Rev. Earth Planet. Sci; 17, 13-53.
18. Giggenbach, W.F., 1988. Geothermal solute equilibria. Derivation of Na-K-Mg-Ca geo indicators .Geochim. Cosmochim.Acta. 52, 2749–2765

Determination the optimal tilt and azimuth angle of photovoltaic system for maximum energy in Bucharest region, Romania – Case Study

Mihaela Andreea Mitiu[1], *Marius Viorel* Olteanu[1],*Gyorgy* Deak[1], *Natalia Simona* Raischi[1], and *Danut* Cociorva[1]

[1]National Institute for Research and Development in Environmental Protection – INCDPM Bucharest, 294, Spl. Independentei, District 6th, Bucharest, Postal Code 060031, Romania

Abstract This study was conducted to evaluate the feasibility of installing a photovoltaic system for supplying the electric load of an environmental institute and to determinate the optimal tilt and azimuth angle of an 100 kW photovoltaic power plant located on the roof of institute. Those parameters plays an important role in maximizing the solar radiation collected by a PV panel. Simulations were made with PVsyst V6.39 for south-east orientation (100; 450) and south-west orientation (100; 450), and for the tilt angle the simulations were made at 100, 150, 370, 600 and 900. The simulation has been done to study the influence of the photovoltaic panels orientation on the amount of electricity produced each month of the year as well as on the total annual amount of electricity. Was experimentally determined by repetitive simulations in the range 300-400 for tilt angle and zero azimuth, optimal parameters for the position of photovoltaic panels. Was defined as the optimal position, the pair of values [zero azimuth, tilt angle 370] where the plant generates maximum energy in one year compared to any other set of values that defines the position of the photovoltaic panels.

1 Introduction

With rapid economic growth and improvement in living standards, there has been a marked increase in energy use for many developed and developing countries[1,2]. Sunlight is by far the largest carbon-free source of energy on the planet. Photovoltaic energy production is becoming more and more applied, being one of the main green energies[3]. The environmental impact due to the emission of pollutants and the consumption of non-renewable resources is increasing, therefore the rational use of resources is necessary and use more and more of viable alternative resources[4,5]. Fossil fuels like oil, coal or gas, are the main sources of energy worldwide and their use causes greenhouse gas emissions into the atmosphere, thereby contributing to global warming[6]. The solar energy is virtually inexhaustible, it is renewable and non-polluting. The production of solar energy by solar panels or other means of using solar energy is noise-free, unlike other methods of producing electricity.

The major and important issue is to improve the efficiency of the photovoltaic power plant, the tilt angle and azimuth angle play important role towards the efficiency of the plant[7,8].Solar panels or PV arrays are most efficient, when they are perpendicular to the sun's rays. It is critical issue to select the both angle to improve the efficiency of a photovoltaic panel [9].The amount of solar radiation captured by a PV depends on the angle of inclination and the orientation angle of the panel. These two parameters are very important for both solar energy experts and designers[10]. The performance of the PV modules varies depending on the location and prevailing environmental conditions to which it is subjected[11]. Because the intensity of the radiation changes at each moment, it is obvious that the production of energy by the solar cells will also be changeable. To achieve the highest PV yield, the local environmental parameters must be taken into account[12].

Optimum tilt angle of photovoltaic panels for different areas of Romania was calculated, the results of the study show that the largest PV plants should be installed in Bucharest, other large PV plants should be built in Constanta, Craiova, Timisoara, Galati, Iasi, Cluj Napoca and Brasov[13].

2 Materials and methods

This energy from the photovoltaic power plant is used to save conventional energy resources. The 100 kW photovoltaic plant contains 400 photovoltaic panels of 1.6m x 1m, polycrystalline cells type and occupies an area of 667m^2.The geographic dates of the location for which the simulations were made are: Bucharest, Romania, latitude 44.5^{0}N, longitude 26.2^{0}E and altitude 81m.At this location exist a weather station MeteoNorm 7.1 that provided annual meteorological weather values for the simulation program. Simulations were made using

PVsyst V6.39 software, for south-east orientation (10^0; 45^0) and south-west orientation (10^0; 45^0), for the tilt angle the simulations were made at 10^0, 15^0, 37^0, 60^0 and 90^0. The software PVsystused in this paper for simulation is a package for the study, sizing and data analysis of complete PV systems. This software can simulate shading and possible energy losses due to shading of a PV. The results are in the form of a full report, specific graphs and tables.

In the table 1 are presented the monthly meteo values for the studied location provided from MeteoNorm 7.1 station and the figure 1 present the solar paths at Bucharest, latitude 44.5^0N, longitude 26.2^0E and altitude 81m.The diagram from figure 1 show the path of the sun across sky from the summer to winter solstice.

Table 1 Monthly meteo Values

	Ian.	Feb.	Mar.	Apr.	Mai	Iun.	Iul.	Aug.	Sep.	Oct.	Nov.	Dec.	An
Hor. global [kWh/m^2.mth]	43.3	69.6	109.9	143.5	181.7	195.8	203.9	178.3	125.0	85.0	45.9	37.7	1415.6
Hor. diffuse [kWh/m^2.mth]	23.6	29.2	46.4	69.7	77.1	79.3	82.9	73.4	55.2	39.5	28.7	20.1	625.1
Extraterrestrial [kWh/m^2.mth]	108.6	142.5	219.6	278.9	337.9	348.2	350.3	310.1	239.4	179.7	117.1	95.1	2728.4
Clearness Index	0.399	0.489	0.500	0.511	0.538	0.562	0.582	0.575	0.522	0.473	0.389	0.365	0.519
Amb. Temper. [^{0}C]	-1.0	1.5	6.7	11.8	17.9	21.1	23.7	23.4	17.1	11.9	6.4	0.4	11.7
Wind velocity [m/s]	2.8	2.9	3.1	3.0	2.6	2.5	2.4	2.3	2.4	2.3	2.6	2.8	2.6

Source MeteoNorm 7.1 station

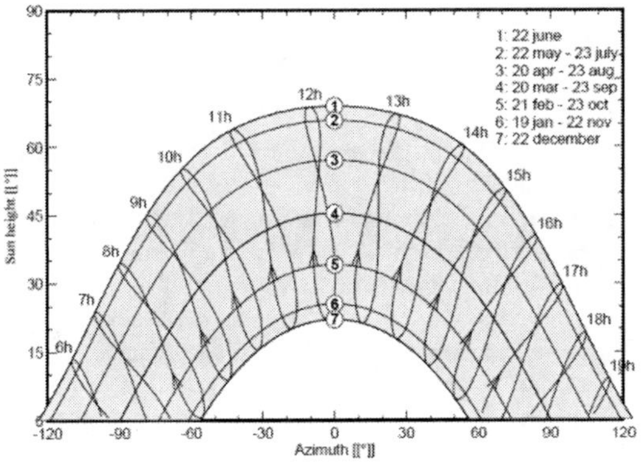

Fig.1 Solar paths at Bucharest (Lat. 44.5^0N, long.26.2^0E. alt. 51m)

To define the orientation of the photovoltaic panels, it is enough to consider two parameters: Azimuth representing the angle of the meridian intersecting the photovoltaic panel plane and the ZERO meridian plus the name of the cardinal point EST or WEST closest to the intersecting meridian. Thus, azimuth = 0 means that the panel is oriented exactly to the SUD. The 10^0 EST azimuth means the panel is ten degrees from the SUD to the EST. And the second parameter is Tilt Angle of the photovoltaic panels to the horizontal plane. This can vary from 0^0, where photovoltaic panels are mounted horizontally up to 90^0, panels mounted vertically.

3 Results and discussions

The simulation on 100 kW photovoltaic power plant located in Bucharest region, Romania has been done to study the influence of the orientation of photovoltaic panels on the amount of electricity produced each month of the year as well as on the total annual amount of electricity. Following the simulations was defined as the optimal position, the pair of values [zero azimuth, tilt angle 370] where the plant generates maximum energy in one year compared to any other set of values that defines the position of the photovoltaic panels. In the table 2 are presented the values obtaining after the simulation for the optimal position of the photovoltaic panels, these data show the energy produced by the 100 kW photovoltaic power plant every month of the year. For the optimal position result after simulation, the total amount of energy that is injected in to the grid for the entire year is 142983kW h.

Table 2. Dates obtaining for values zero azimuth, tilt angle 370 for the photovoltaic power plant

	Gl. Horiz. [kWh/m^2.day]	Coll.Plane [kWh/m^2.day]	System output [kWh/day]	System output [kWh/month]
Jan.	1.40	2.57	217.8	6753
Feb.	2.49	4.12	349.9	9796
Mar.	3.55	4.85	411.7	12764
Apr.	4.75	5.35	454.2	13626
May	5.86	5.87	498.1	15442

 https://doi.org/10.1051/e3sconf/20198001007

June	6.53	6.16	523.0	15691
July	6.58	6.34	538.5	16693
Aug.	5.75	6.15	522.2	16187
Sep.	4.17	5.21	442.6	13277
Oct.	2.74	4.12	350.1	10853
Nov.	1.53	2.50	212.2	6367
Dec.	1.12	2.10	178.5	5534
Year	3.88	4.61	391.7	142983

In figure 2 are presented the evolution of the meteo and incident energy and system output after simulation, for the optimal position of the photovoltaic panels, the data are for the 100 kW photovoltaic power plant located in the area of Bucharest.

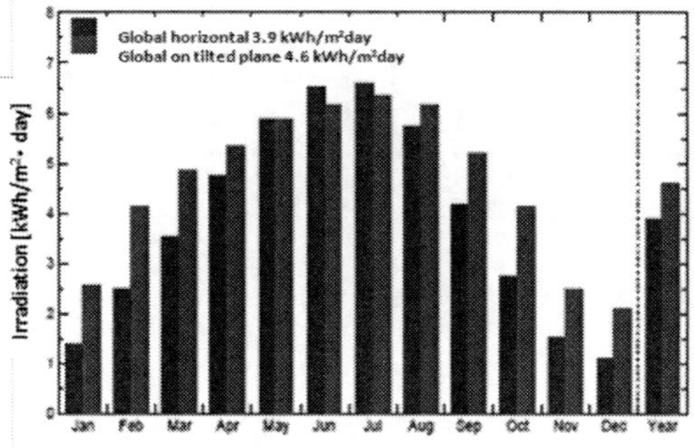

(a)

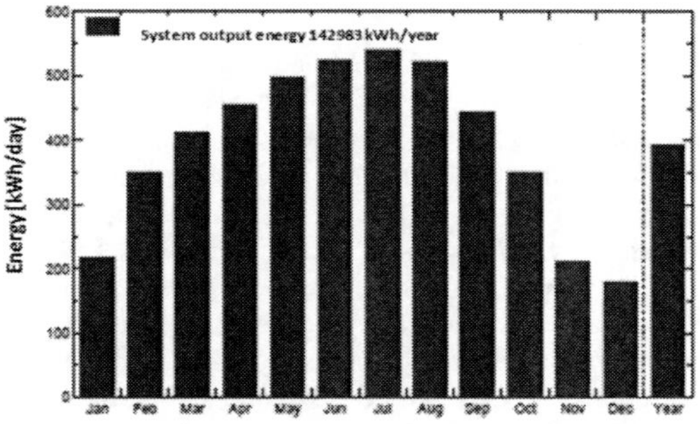

(b)

Fig.2 Meteo and incident energy (a) and system output(b), for the optimal position of the photovoltaic power plant

In July the 100 kW photovoltaic power plant generated the maximum energy, 16693 kWh (a daily average of 538.5 kWh), ie 11.67% of the total energy produced in one year. The minimum energy generated by the power plant was in December, 5534 kWh, with a daily average of 134.3 kWh. This represents 33.15% of the energy produced in July and 3.87% of the total energy generated in one year (142983kWh).

In order to make it easier to observe the energy gain according to the inclination angle of the photovoltaic panels with respect to the horizontal plane, it can vary from 0^0,where the photovoltaic panels are horizontally mounted, up to 90^0, panels mounted vertically and South orientation, the energies were compared each month for each tilt, that is presented in the figure 3.

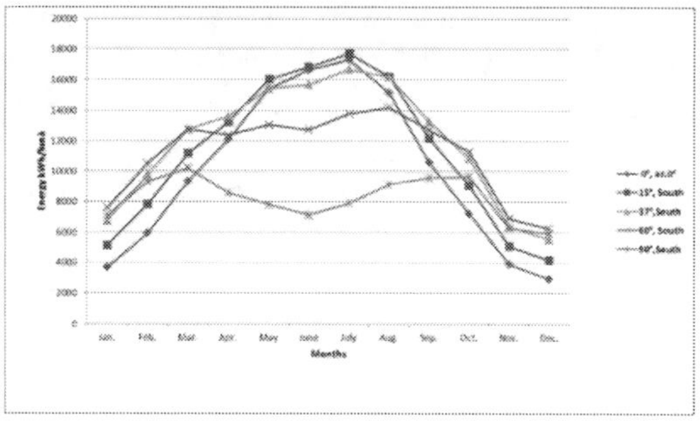

Fig.3 Monthly energy depending on tilt angle and south azimuth for the photovoltaic power plant located in the area of Bucharest

Curve 37^0- South from the figure 3closes the largest surface / area. This means that the whole year's energy production is the highest and is therefore the optimal position. Compared with the position at $60^0$where the latter has a higher production starting with the autumn equinox and up to the spring equinox, overall annual production is 6% lower at the 60^0.In the case of vertically mounted ones, they have a higher production 15-20 days before and after the winter solstice, but throughout the year have a lower production by 31% than the optimal position of 37^0.

Likewise, the panels mounted at 15^0-South, from May until August, produce more energy than those mounted in the optimal position at 37^0. But throughout the year, those inclined at 15^0 have a 6% loss compared to the optimal inclination. However, if we look from the point of view of an investor on the Black Sea coast who wants to mount photovoltaic panels on the building, in the summer season he would have a higher gain with the 15^0inclination. For him the optimum is at 15^0 where the consumption curve is higher in summer.

The energy values obtained (fig.4) for the same angle of inclination and azimuth were compared to see the optimal position where we have the highest energy gain.

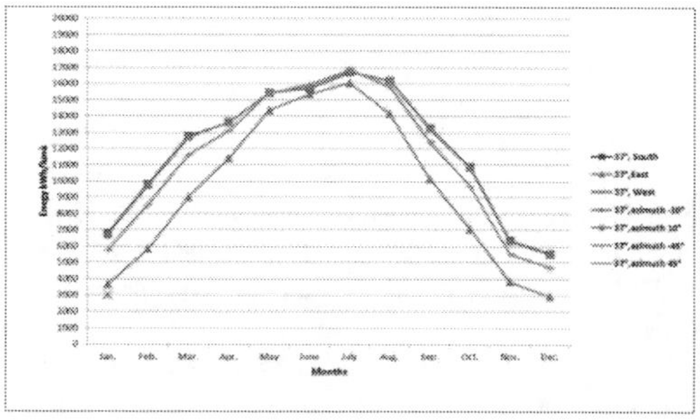

https://doi.org/10.1051/e3sconf/20198001007

Fig.4 Monthly energy depending on different azimuth and the same tilt angle for the photovoltaic power plant located in the area of Bucharest

From figure 3 we can easily see how the curves at 37^0- azimuth 0^0, 37^0-azimuth + 10^0, 37^0-azimuth-$10^0$are very close together, almost overlapping. This denotes that a + 10^0 deviation (even +15^0) does not affect the energy gain of the plant. At a deviation of + 45^0, the loss already counts, and at +90^0 is significant. In the case when the roof has two plans, one to the East and the other to the West it will be mount half of the photovoltaic panels to the East and the other half to the West. In this way the losses are eliminated.

4 Conclusions

Following the simulations performed with the PVsyst V6.39 program, on 100 kW photovoltaic power plant, consisting of 400 PV, stretched over a surface of 667 m^2, with dimensions 1.6mx1m, of polycrystalline type, resulted in the following conclusions:

The optimal position of the photovoltaic panels regarding on the annual energy gain is the one at the 37^0 tilt angle and the azimuth 0^0. However, in this position, in the summer months, was not recorded the highest energy gain. On the total of the 12 months of the year is the position where the plant generates the largest amount of energy.

A + 10^0-15^0 deviation from the optimal position (South orientation) does not greatly influence the energy gain. As shown by the values resulting from the simulation, the losses in this case are 0.3%, insignificant. At 45^0 the energy gain decreases by 5.5% compared to the optimal position.

The total East and West orientation reduces energy gain by 20.5% respectively 19.9%, so we can consider a 20% loss in both cases. Panels mounted at 15^0South, starting with May until beginning of August, produce more energy than those mounted in the optimal position at 37^0. But all year round, those inclined at 15^0 have a loss of 6% compared to the optimal inclination.

At a deviation of +45^0 already the loss counts, and at + 90^0 it is significant. If the roof has two plans, one to the East and the other to the West, will mount half of the photovoltaic panels to the East and the other half to the West. In this way, the losses are eliminated.

The research activities that underlie this article has been performed in a project which is part of Research Programe NUCLEU - contract 48N/2016 (PN 16 04 02 01), financed by the Romanian Ministry of Research. The authors would like to thank the management and employees of the National Institute for Research & Development in Environmental Protection for their valuable assistance and suggestions.

References

1. H.W. Li Danny and N. T. Lam Tony, International *Journal of Photoenergy*, 9 pages, (2007)
2. Mian Guo, Haixiang Zang, Shengyu Gao, Tingji Chen, Jing Xiao, Lexiang Cheng, Zhinong Wei 1 and Guoqiang Sun *Appl. Sci.*7,10, (2017)
3. Qingguo Xu, Hongbo Li, Guoqiang Hao, and Yefei Ding, *JOCET* **4**, 6, (2016)
4. E. Holban, GY. Deák, V. Daescu, et al., *JEPE* **16**, 2, 479–488, (2015)
5. GY. Deak, V. Daescu, E. Holban, et al., *JEPE* **16**, 1, 304–315, (2015)
6. Ma J., Man K. L., Ting T. O., Zhang E. G., Guan S., Wong P. W., Krilavicius T., Sauleviius D., and Lei C. U. , *ktu*, **20**,1, 41-44, (2014)
7. Radhika, S.K. Suman (2015), *IJAREEIE*, **4** 6, (2015)
8. M. Hartner, A. Ortner, A. Hiesl, R. Haas, *Appl.* Energ. **160**, 94–107, (2015)
9. D. Jain, M. Lalwani, *IJRER*, **7**, 3, 1053-1061, (2017)
10. K. Markam, K. Sudhakar, *ITJET*, **3**, 5, 2735 – 2741, (2016)
11. E. B. Ettah, E. E. Eno, and A. B. Udoimuk, *Journal of* Association *of Radiographers of Nigeria*, **23**, 16–22, (2009)
12. A. R. Gxasheka, E. E. van Dyk, and E. L. Meyer, *Renewable Energy*, **30**, 4, 611–620, (2005)
13. V. Badescu, F. Iacobescu (2013), *Energy for* Sustainable *Development* **17**, 220–227, (2013)

Development of a footstep power generator in converting kinetic energy to electricity

Chun Kit Ang[1], *Ammar A.* Al-Talib[1], *Sook Meng* Tai[1], and *Wei Hong* Lim[1]

[1]Department of Mechanical and Mechatronic Engineering, Faculty of Engineering, Technology and Built Environment, UCSI University, 56000 Kuala Lumpur, Malaysia

Abstract. Kinetic energy is considered as one of the renewable energies. Substantial amounts of researches were conducted to investigate the feasibility of converting the kinetic energy into electricity. Nevertheless, most of these previous works emphasized on the selection of suitable materials and the design of power generator which tends to be complicated. In this paper, a simple and yet low cost mechanism has been proposed to enhance the performance and efficiency of energy conversion from kinetic energy to electricity energy by placing a mechanical footstep power generator on the hind foot region. A total of 45 individuals were invited to participate in the experiments and the experiment results are then compared with the theoretical results.

1 Introduction

The deployment of different clean energy systems is a crucial strategy to achieve environment sustainability [1-4]. Most people are spending most of their lifetime in walking. Walking is also known as ambulation, i.e., a fundamental and common locomotion for human in daily life. The contacts between human feet and ground surface is created during the walking. The forces experienced by human feet upon landing on the ground can generate a renewable energy known as kinetic energy. This energy can be converted into electricity through a footstep power generator.

Different types of footstep power generators are available in the market and majority of these devices use piezoelectric transducer to generate power. One of the greatest challenges in designing the footstep power generators with piezoelectric transducer is the selection of suitable ferroelectric material because it governs the efficiency of converting kinetic energy to electricity [5]. Conventionally, the piezoelectric footstep power generator uses ferroelectric materials made up of crystal such as Lead (II) titanate ($PbTiO_3$), Lead (II) Zirconate ($PbZrO_3$), Polyvinylidene Difluorideor Polyvinylidene Fluoride (PVDF) and Lead Zirconate titivate (PZT). While both of the PZT and Polyvinylidene Difluorideor (PVF) are commonly used as piezoelectric, the former material is the best candidate of piezoelectric because it can produce better output voltage than other ferroelectric materials [6]. Despite of its popularity, piezoelectric footstep power generator suffers with some drawbacks such as the infeasibility of this technology under static condition and the limitation of power capacity [7].

Mechanical footstep power generator is an alternative of footstep power generator that can produce electricity with different mechanisms. Instead of using piezoelectric transducer, delicate mechanisms are designed in the mechanical footstep power generator to convert human motion into electricity. Some notable advantages offered by the mechanical footstep power generator include the simplicity of its assembly process, lower cost, lesser transmission lost and etc. [8].

There are many types of combination for mechanical footstep power generator in the market. For example, rack and pinions with pulley [9], crank shaft with chain drive system [10], fly wheel with gear [11] and etc. However, most of these designs are combined with two mechanism components such as rack and pinion with fly wheel, rack pinion with pulley systems, and rack and pinion with chain drive system.

In this research, a mechanical footstep power generator is designed based on the concept of rack and pinion in order to reduce the complexity of mechanical structure. The components included in the proposed footstep power generator are rack and pinion, shaft, spur gears, bearings, DC generator and battery. The rest of this paper are presented as follows. Section 2 explains the design and fabrication of mechanical footstep power generator. Sections 3 and 4 conducted experiments and discussed the results, respectively. Finally, Section 4 concludes the works of this paper.

2 Design and fabrication

As mentioned in the earlier section, the mechanical footstep power generator in this research is designed based on the rack and pinion concept, aiming to reduce the complexity of mechanical structure. Figure 1 illustrates the design of mechanical footstep power generator.

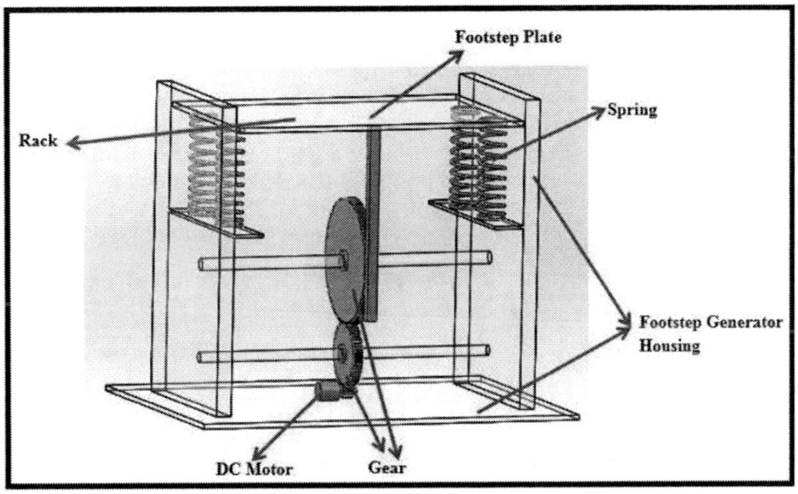

Fig. 1. Proposed mechanical footstep power generator.

Table 1. Ideal specification of major mechanical components in the proposed mechanical footstep power generator.

	Unit	Gear 1 (First Gear)	Gear 2 (Second Gear)	Gear 3 (Last Gear)
Face Width	mm	10	10	10
Base Circle	mm	112.77	56.38	11.25
Addendum	mm	122	62	14
Teeth	-	120	60	12
		Spring		
Wire Diameter	mm	2		
Outer Diameter	mm	44		
Inner Diameter	mm	40		
Total number of coil	Turns	65.69		
Free Length	mm	231.38		
Spring Stiffness	N/mm	4.09		

As shown in Figure 1, the rack and pinion drive the first gear due to the planter force exerted on the footstep plate. The second gear that is attached to a DC generator or dynamo will then be driven by the first gear. The last gear attached with DC generator will eventually transfer the kinetic energy into electricity and store the output in battery through the electronic circuit. The specification of each major component of mechanical footstep power generated is presented in Table 1. These specifications are given based on the design illustrated in Figure 1 and they are considered as the ideal values to be used for the calculation of theoretical results.

3 Experiment setup and data collection

The test subjects in this study involved of nine groups of healthy volunteers with each group consists of five individuals without any foot disorders. The participants in each group can either be kids or adults with the body weights range between 15 kg to 55 kg. The average reading from each group was taken in order to minimize the variation. The test subjects were instructed to use their hind foot to step on the proposed mechanical footstep generator because previous studies done by [12] showed that the hind foot exerts most pressure, i.e., nearly one third of the entire foot pressure, when walking on the flat plane. Furthermore, the previous studies of [13] reported that no significant differences were detected when comparing plantar pressure distribution produced by both foot of men and women on flat surfaces, implying that the negligible impacts of gender factor [13]. Based on these findings, this research assumes that the power generated from each individual can be maximized by using hid foot to step on the proposed mechanical footstep generator.

In this research, different LEDs were used to indicate the intensity of power generated by the proposed mechanical footstep power generator because different ranges of voltage are required for lighting up the LEDs with different colors as shown in Table 2. The generated power can be calculated through the measured values of LED forward voltage and the current flew through it.

Table 2. Voltages required for LEDs with different colors.

Color	Voltage Required (V)
Red	1.6 – 2.0
Green	1.9 – 4.0
Yellow	2.1 – 2.2
Blue	2.5 – 3.7
White	3.1 – 4.4

4 Result and discussion

As compared to the theoretical results using specification in Table 1, the actual results obtained in Table 3 were slightly different. This is because the gears and spring were bought and modified as shown in Table 3, leading to the deviations.

In terms of dimensional sizes, the actual sizes of gear 1, gear 2 and gear 3 are 4.09%, 1.95% and 2.22% larger than the associated ideal values. Nevertheless, the overall performance of the proposed mechanical footstep power generator is not significantly impacted because the tolerance is less than 5% for each gear. As for the stiffness of spring, the difference is up to 13% because the materials of actual spring chosen for this project is STWC80C steel with 80% of carbon, while the materials chosen for design at the beginning stage was oil temper wire. These differences of materials used in the springs are identified as the main factor that contribute to the deviations between theoretical and actual results.

The differences between the theoretical and actual results for different ranges of mass of individual are presented in Table 4 and Figure 2. It is observed that the

Table 4. Differences between the theoretical and experimental results.

Mass (kg)	Theoretical Power (W)	Experimental Power (W)
15-19	0.0736	0.0603
20-24	0.0981	0.0801
25-29	0.1226	0.1011
30-34	0.1472	0.1390
35-39	0.1717	0.1504
40-44	0.1962	0.1830
45-49	0.2207	0.2012
50-54	0.2453	0.2301
55-59	0.2698	0.2900

Table 3. Actual specification of major mechanical components in the proposed mechanical footstep power generator.

	Unit	Gear 1 (First Gear)	Gear 2 (Second Gear)	Gear 3 (Last Gear)
Face Width	mm	10	10	10
Base Circle	mm	117.5	57.5	11
Addendum	mm	122	62	14
Teeth	-	120	60	12
		Spring		
Wire Diameter	mm	2.95		
Outer Diameter	mm	29.46		
Inner Diameter	mm	23.56		
Total number of coil	Turns	12.75		
Free Length	mm	100		
Spring Stiffness	N/mm	3.53		

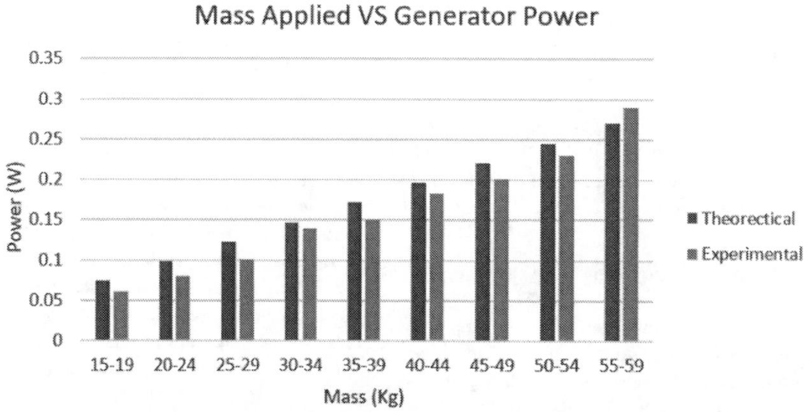

Fig. 2. Differences between theoretical and experimental results.

generated power increased proportionally with the mass of an individual. In general, the individual with body mass around 55 kg are the teenager or adults with age above 18 years old. These participants are normally stronger, hence the actual power generated from experiments were higher than the theoretical results. Furthermore, it was also observed that the spring with higher stiffness needs to be installed to the mechanical footstep generator designed for adults to ensure it has higher durability.

5 Conclusion

The conversion of kinetic energy from human footsteps into the electricity energy is considered as one of the renewable energy techniques. This research had demonstrated the feasibility of generating electricity with simple mechanism of rack and pinion. Furthermore, this research proposed to place the mechanical footstep power generator at hind foot region in order to generate higher output power with better efficiency. Experiments were conducted to verify the effectiveness of the proposed design and the experimental findings observed are promising.

References

1. E.C Chang, Y.C Liu, Int. J. Smart Grid Clean Energy **7**, 1 (2018)
2. J. Thakur, S. Rauner, B. Chakraborty, Int. J. Smart Grid Clean Energy, **6**, 3 (2017)
3. Y. Stefanov, K. Ivanov, P. Petrov, Int. J. Smart Grid Clean Energy, **6**, 3 (2017)
4. J. Zhang, A. Daniela, Y. Fang, A. Desmonda, E. O. Antwi, Int. J. Smart Grid Clean Energy, **7**, 3 (2018)
5. H.R. Nyan, J. Material Sci. Eng. **4**, 3 (2015)
6. S.B. Mohanty, S.S. Panda, Int. J. Eng. Sci. Res. Tech. **3**, 5 (2014)
7. V. Mathane Nitashree, L. Salunkhe Arati, S. Gaikwad Sayali, Int. J. Adv. Res. Elec. Comm. Eng. **4**, 10 (2015)
8. V. Tom Jose, B. Boban, M.T. Sijo, Int. J. Sci. Res. Pub. **3**, 3 (2013)
9. A. Farouk, A. A. Gawad, RTES, (2015)
10. N. Modi, P. Shrivastava, R. Bhardwaj, U. Jaiswal, Int. Res. J. Eng. Tech. **3**, 5 (2016)
11. A. Muhammad, A. Ali, S. Bibi, Ener. Environ. Sus. Dev. **275** (2016)
12. E. Klimiec, K. Zaraska, J. Piekarski, P. Guzdek, G. Kolaszczynski, B. Jasiewicz, Sens. Actuator A-Phys. **247** (2016)
13. R. Periyasamy, A. Mishra, S. Anand, A.C. Ammini, Foot. **21**, 3 (2011)

Experimental research of a four-stroke piston type compressed air engine

Fancong Zeng[1], *Jinli* Xu[1], and *Rui* Zhao[1]

[1]School of Mechanical and Electrical Engineering, Wuhan University of Technology, Wuhan, China

Abstract. The purpose of this paper is to investigate working performance for a four-stroke piston type compressed air engine (CAE) via experimental research. A conventional internal combustion engine(IC engine) is modified to a four-stroke CAE and a mathematical model is set up. Working performances are investigated by output torque, power, air consumption, exhaust pressure, exhaust temperature and cylinder pressure on a test bench. The CAE is installed on a vehicle for the road test. The experiment and analysis show several meaningful results. The increase of the supply pressure will result in the increase of the rotational speed, and it can effectively improve the output torque and power. However, it can lead to an increase in air consumption. The success of this application demonstrates compressed air cars are likely to be industrialized duo to their zero pollution and zero emissions.

1 Introduction

Over the recent decades, environmental protection and reducing carbon dioxide (CO_2) emission become crucial issues worldwide. Exhaust gas from internal combustion (IC) engines contributes largely to CO_2 emissions, and IC engines pollute the environment seriously. In order to improve fuel consumption and protect the environment, researchers have been searching for environmentally friendly energies to replace conventional fossil fuels [1, 2]. Compressed air, as the working medium of the compressed air engine (CAE), is inexhaustible. A CAE has both the advantages of non-burning fuel and zero pollution, and there is no chemical reaction and high temperature in its working cycle [3, 4]. Compressed air can be generated through several ways, such as solar energy, wind energy and tidal energy [5], which does not rely on non-renewable energy.

Lots of literatures on the CAE have been investigated. Zhang et al. [6] studied the valve timing for optimizing the efficiency and performance of the CAE. An optimized polynomial cam profile was applied and the influences of the half wrap angle and the maximum valve lift to the fullness coefficient were studied. Song et al. [7] established the mathematic model and dimensionless of the CAE, and the influences of the non-dimensional cylinder stroke and Kagawa coefficient on CAE's performances were analyzed. The research results showed that the non-dimensional cylinder stroke had little effect on the CAE, but the heat transfer has a remarkable effect on the performance of the CAE. The faster the heat transfer was, the larger the efficiency and power were. Huang et al. [8] presented an experimental investigation of a CAE. The CAE was a modified 100cm^3 two-stroke IC engine obtained from a motorcycle manufacturer. The experiments in this study used a test bench to examine the power performance and pressure/temperature variations of the CAE at pressures ranging from 5 to 9 bar.

In this paper, an experimental research of a four-stroke piston type compressed air engine was carried out, the influences on the output torque, power, air consumption, exhaust pressure, exhaust temperature and cylinder pressure were analyzed with the variations in rotational speed and supply pressure. Therefore, the experimental research of the four-stroke CAE is carried out in this paper for performance analysis and the design of the CAE.

2 Mathematical model of the CAE

As shown in Figure. 1, inlet valves and exhaust valves are controlled by the valve system. At the beginning of the intake stroke, the intake valves open immediately, and the atmosphere flows into the cylinder through the inlet valves because of the pressure difference. When the piston is near the bottom dead center (BDC), the inlet valves close. In the compression stroke, the piston moves up near the top dead center (TDC). In the inflation and expansion stroke, the intake valve and the air nozzle begin to work, then the compressed air flows into the cylinder through the air nozzle, and pushes the piston downwards. After a specific crank angle, the intake valve and the air nozzle close while the compressed air expands to push the piston down and output work. When the piston moves back to the bottom dead center (BDC), the exhaust valves open so that residual air discharges under the impetus of the piston. After the piston moves back to the top dead center (TDC), the four-stroke CAE completes a complete work cycle.

The CAE cylinder can be seen to be a thermodynamic system, and the following assumptions are made. First, the compressed air is ideal, which means specific heat and specific enthalpy are only related to the temperature. Second, the changes of kinetic and gravitational energy are negligible. Third, the process of gas flowing into/out of the cylinder is a quasi-steady one. Finally, there is no leak or friction during the working process.

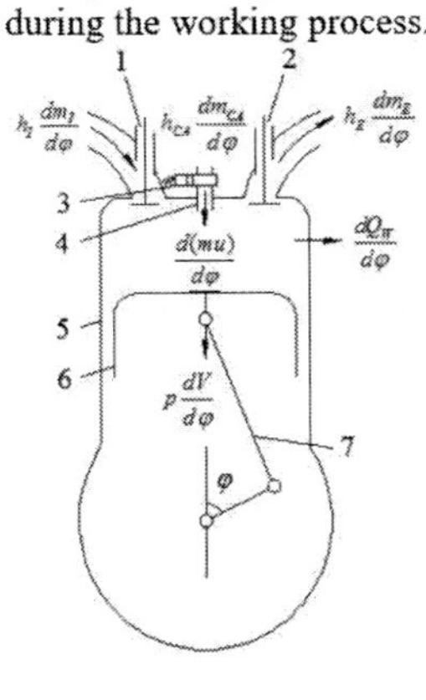

1. Inlet valve 2. Exhaust valve 3. Intake valve 4. Air nozzle 5. Cylinder 6. Piston 7. Crank-link mechanism

Fig. 1. Structure and thermodynamic analysis diagram of four-stroke CAE

The energy equation can be expressed as

$$\frac{dU}{d\varphi}=\frac{dQ_w}{d\varphi}+h_I\frac{dm_I}{d\varphi}+h_{CA}\frac{dm_{CA}}{d\varphi}+h_E\frac{dm_E}{d\varphi}+\frac{dW}{d\varphi} \quad (1)$$

Where U is the internal energy of the air in the cylinder; Q_w is the heat absorbed by the air from the cylinder walls; h_j (j=I, E, CA) are the specific enthalpies of inlet air, exhaust air and injection compressed air, respectively; m_j (j=I, E, CA) are the mass of inlet air, exhaust air and injection compressed air, respectively; W is the mechanical work on the piston; φ is the crank angle.

U can be written in this form

$$dU=d(m\cdot u)=udm+mdu \quad (2)$$

Where m is the mass of the air in the cylinder; u is the specific heat.

For ideal air, u can be seen as a function of air temperature, that is u=$u(T)$, and u can be written in a fully differential form

$$\frac{du}{d\varphi}=\frac{\partial u}{\partial T}\frac{dT}{d\varphi}=c_v\frac{dT}{d\varphi} \quad (3)$$

Where c_v is the constant volume specific heat; T is the temperature of the air in the cylinder.

The heat transfer in the process can be expressed as

$$\frac{dQ_w}{d\varphi}=\frac{1}{\omega}\sum_{i=1}^{3}a_g\cdot A_i(T_{wi}-T) \quad (4)$$

Where ω is the angular speed of the crank shaft; a_g is the heat transfer coefficient; A_i is the total heat transfer area; T_{wi} (i=1, 2, 3)are the temperatures of the cylinder head, the piston and the cylinder liner, respectively.

The coefficient of the heat transfer between the air and the cylinder walls is given as

$$a_g=0.1129\cdot D^{-0.2}\cdot p^{0.8}\cdot(Sn/30)^{0.8}\cdot T^{-0.594} \quad (5)$$

Where D is the diameter of the cylinder; p is the pressure of the air in the cylinder; S is the stroke of the piston; n is the rotational speed of the CAE.

The mechanical work by the compressed air is described by

$$\frac{dW}{d\varphi}=-p\frac{dV}{d\varphi} \quad (6)$$

Where V is the cylinder volume, V can be written as

$$V=\frac{\pi D^2}{4}\{\frac{S}{\varepsilon-1}+\frac{S}{2}[(1+\frac{1}{\lambda}) -\cos\varphi-\frac{1}{\lambda}\sqrt{1-\lambda^2\cdot\sin^2\varphi}]\} \quad (7)$$

Where ε is the compression ratio; λ is the ratio of crank radius to connecting rod.

The change rate of the cylinder volume can be expressed as

$$\frac{dV}{d\varphi}=\frac{\pi\cdot D^2\cdot S}{8}[\sin\varphi+\frac{\lambda}{2}\frac{\sin(2\cdot\varphi)}{\sqrt{1-\lambda^2\sin^2\varphi}}] \quad (8)$$

Substituting equations (2)–(6) into equation (1) yields the following differential equation

$$\frac{dT}{d\varphi}=\frac{1}{m\cdot c_v}(\frac{dQ_w}{d\varphi}-p\frac{dV}{d\varphi}+h_I\cdot\frac{dm_I}{d\varphi} +h_{CA}\cdot\frac{dm_{CA}}{d\varphi}+h_E\cdot\frac{dm_E}{d\varphi}-u\frac{dm}{d\varphi}) \quad (9)$$

The mass air flow of the CAE can be expressed as

$$\frac{dm_j}{d\varphi}=\frac{1}{\omega}\cdot\mu F\sqrt{\frac{p_1}{v_1}}\cdot\psi \quad (10)$$

Where μ is the coefficient of discharge; F is the effective sectional area; p_1 is the upstream pressure; v_1 is the upstream air specific volume; ψ is the flow function.

When $\frac{p_r}{p_1}>(\frac{2}{k+1})^{\frac{k}{k-1}}$, the air flow state is sub-sonic flow state, and the flow function ψ can be written as

$$\psi=\sqrt{\frac{2k}{k-1}[(\frac{p_r}{p_1})^{\frac{2}{k}}-(\frac{p_r}{p_1})^{\frac{k+1}{k}}]} \quad (11)$$

Where k is the adiabatic exponent of the air; p_r is the downstream pressure.

 https://doi.org/10.1051/e3sconf/20198002002

When $\frac{p_r}{p_1} \le (\frac{2}{k+1})^{\frac{k}{k-1}}$, the air flow state is sonic flow state, and the flow function ψ can be written as

$$\psi = (\frac{2}{k+1})^{\frac{1}{k-1}} \sqrt{\frac{2k}{k+1}} \tag{12}$$

The ideal gas equation of state can be derived as

$$pV = mRT \tag{13}$$

Here, R is the gas constant of air.

3 Experimental method

In this study, the type of the CAE is a four-stroke piston type CAE, which is modified from a four-stroke IC engine. It can output the mechanical energy transformed from the compressed air energy into the torque form. The four-stroke CAE system is mainly composed of main air supply system, branch air supply system, electronic control system and CAE body. Figure 2 is the physical map of the four-stroke CAE system. The main air supply system is connected with air supply pipes through its main flow meter, which is combined with the branch air supply system. The main air supply system stores high-pressure compressed air prepared by an air compressor in air tanks, which acts as the power medium of the CAE. The branch air supply system fills the compressed air into four cylinders, respectively, and the compressed air is obtained through the high-pressure and large-flow pressure reducing valve and controlled by four intake valves. The electronic control system is divided into two parts: ECU electronic control system and bench test system. ECU electronic control system mainly refers to the control system of ECU development platform of the CAE, and also involves the control of timing and pulse width of the intake valves, and can realize the data acquisition and real-time display of the air pressure and air flow of the CAE. The platform test system can collect and measure the cylinder pressure, the exhaust pressure and the exhaust temperature when the CAE is working, and the output torque and power can also be measured.

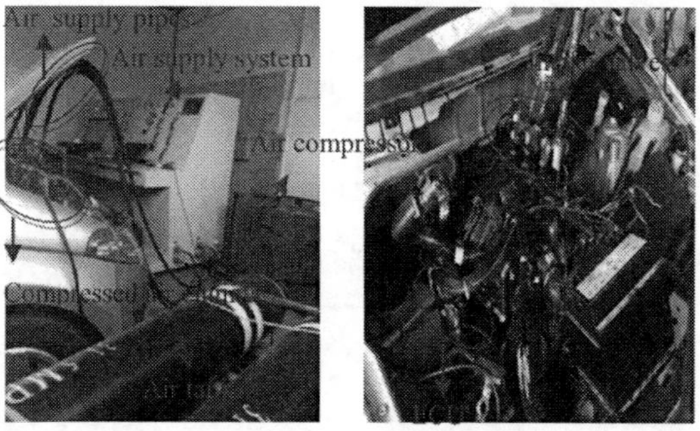

Fig. 2. Physical map of the four-stroke CAE system.

Fig. 3. CAE test bench.

In order to verify the performance of the four-stroke CAE, calibration software and data acquisition system are used to test the CAE on the test bench, the CAE test bench is shown in figure 3 and the measurement and control system is shown in figure 4. The debugged four-stroke CAE is installed on the vehicle for the road test to test the operation of the compressed air car under different working conditions. Figure 5 shows the compressed air car developed in this study. The following conclusions can be obtained through the road test: (1) Vibration and noise of the CAE are much less than that of traditional IC engine, and the operation stability is well. (2) The prototype car can run easily when it is in first gear, the supply pressure required in the initial stage is bigger than that in the process of driving. At the same time, driving speed can be significantly increased by increasing the supply pressure and reducing the weight of the car. (3) Since the prototype car is a rear wheel drive car and it is very heavy, the CAE will shake after shifting gears, the rotational speed can't reach higher speed range.

Fig. 4. Measurement and control system.

Fig. 5. Compressed air car.

4 Result and discussion

Working characteristics of the four-stroke CAE can be investigated by output torque, power, air consumption, exhaust pressure, exhaust temperature and cylinder pressure. Taking the rotational speed as the independent variable, the variation of each important performance parameter with the rotational speed is discussed in this paper. The supply pressure is set to 20bar and 30bar, respectively, and the resistance torque is set to 2Nm, the performance parameters are recorded at different rotational speeds and the test results are shown in figure 6~figure 9. From these figures, it can be found that the higher the rotational speed, the smaller the output torque but the larger the power and air consumption. When running at a lower rotational speed, it consumes less air and has better economic performance. The exhaust pressure and the exhaust temperature gradually increase with the rising of rotational speed for the same supply pressure. With the increasing of output torque, the exhaust temperature will become lower, even below zero. The larger the output torque, the faster the exhaust temperature decreases. The increase of supply pressure will result in the increase of output torque, power and air consumption and the decrease of exhaust temperature.

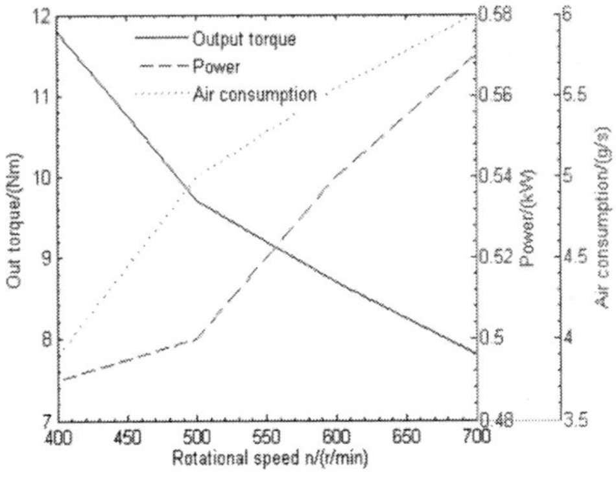

Fig. 6. Variations of output torque, power and air consumption at different rotational speeds. (20bar)

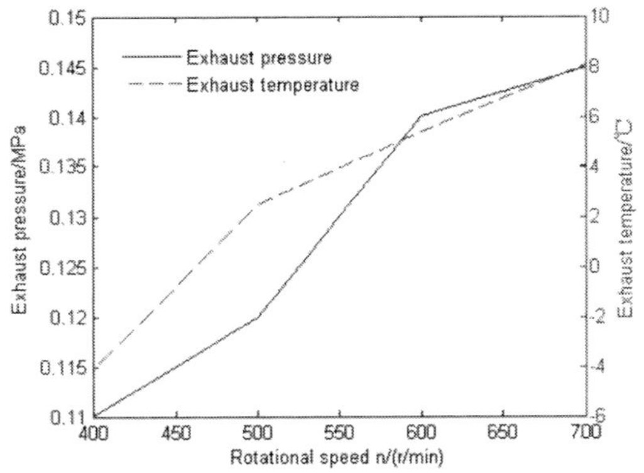

Fig. 7. Variations of exhaust pressure and exhaust temperature at different rotational speeds. (20bar)

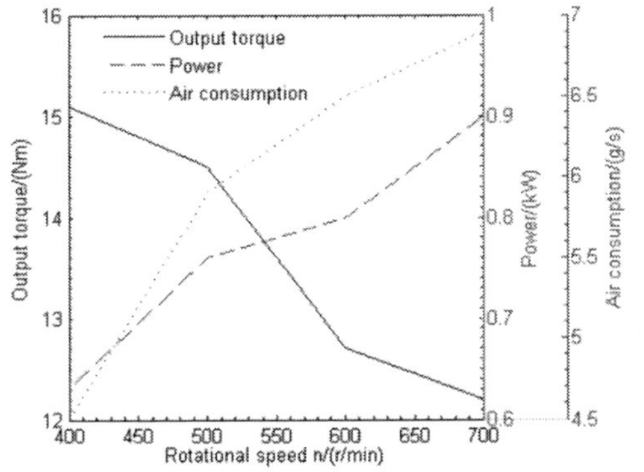

Fig. 8. Variations of output torque, power and air consumption at different rotational speeds. (30bar)

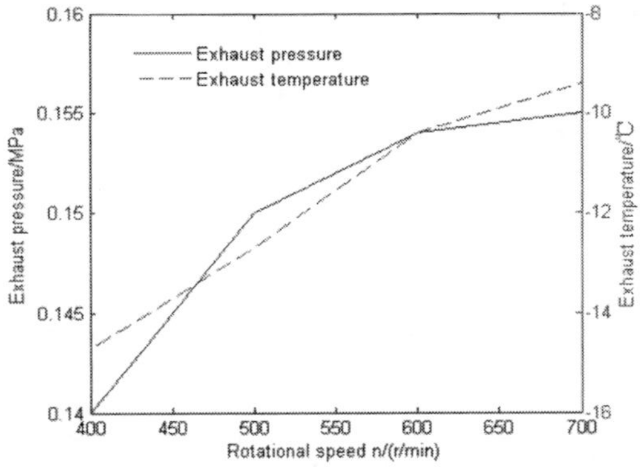

Fig. 9. Variations of exhaust pressure and exhaust temperature at different rotational speeds. (30bar)

As shown in figure 10, the supply pressure is set to 20bar and 30bar, respectively, and the variations of cylinder pressure in a working cycle are revealed. The cylinder pressure increases rapidly in the compression stage (180°CA~355°CA), and the peak cylinder pressures are about 16.5bar and 25bar, respectively, the cylinder pressure drops rapidly in the expansion stage (405°CA~530°CA).

https://doi.org/10.1051/e3sconf/20198002002

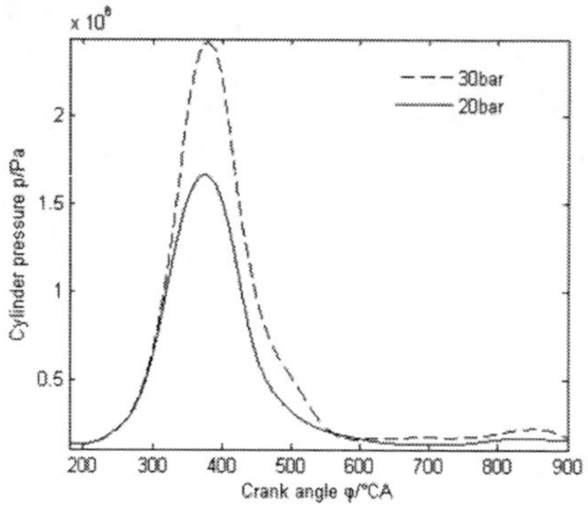

Fig. 10. Cylinder pressure plot.

The peak cylinder pressure is lower than the supply pressure, mainly due to the following reasons:

(1) Although the supply pressure is certain during the test, the air flow rate is smaller than that of the theory due to the limitation of the air flow rate of the intake valves.

(2) The compressed air diffuses gradually in the cylinder, and the cylinder pressure sensors are installed far away from the center of the air-filled flow field, the measured cylinder pressure will be lower than the peak cylinder pressure.

(3) Due to the delays in the opening and closing of the intake valves, and there will be air throttling in the pipeline from the outlets of the intake valves to the jets of the CAE. The piston is already in the downward process when the air is filled into the cylinder, and the cylinder volume gradually increases. At this point, the flow of insufficient compressed air into the cylinder will reduce the cylinder pressure.

(4) Although relevant seal processing has been done in the modification of the CAE, there is still gap between piston ring and cylinder diameter surface. The mounting holes of the valves and cylinder pressure sensors are not completely sealed either, these are the channels of air leakage, which directly lead to insufficient cylinder pressure.

In conclusion, when the air flow is larger, the time of establishing cylinder pressure will be shorter and the peak pressure will be greater. Therefore, the improvement and optimization of the four-stroke CAE in cylinder and valve will be the focus of future research.

Driving torque and resistance torque of the CAE are balanced during stable operation. By comparing and analyzing the variation of the rotational speed with the supply pressure when the CAE is under no load and 2Nm load, the characteristic of the resistance torque is investigated, thus reflecting the power performance. The variations of rotational speed at different supply pressures are shown in figure 11. As observed from this figure, the rotational speed at no load is obviously higher than that at 2Nm load under a certain supply pressure, and the higher the supply pressure, the greater the difference between the rotational speeds of no load and 2Nm load. This indicates that the CAE needs to consume more pressure energy when overcoming the load to output mechanical work, and the rotational speed and power can be improved by increasing the supply pressure and air flow rate of compressed air.

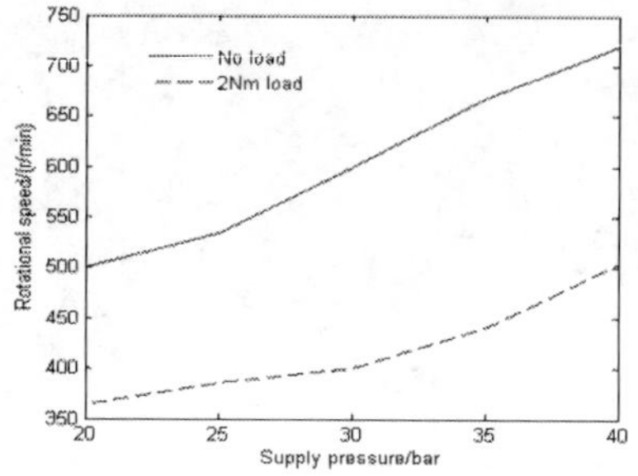

Fig. 11. Variation of rotational speed at different supply pressures.

5 Conclusions

In this paper, a four-stroke piston type CAE is used to investigate the working performance. An experimental system is developed to research the influence of several key performance parameters on working characteristics, and the success of car test drive with four-stroke CAE demonstrates the feasibility of vehicle applications. The CAE has higher output torque, lower power and air consumption in lower rotational speed stage, and the exhaust pressure and the exhaust temperature gradually increase with the rising of rotational speed. The increase of the supply pressure can effectively improve power performance. However, compressed air car has limited range and power, due to the low energy density of compressed air. Therefore, there are still many problems to be solved in the application of the CAE as main power system on vehicles, and the study of this work will provide a certain guidance function for further study on optimizing design of the CAE.

The authors would like to thank the automobile company SGMW in China for its kind assistance with the four-stroke IC engine fused for the present study.

References

1. X.L. Yuan, X. Liu, J. Zuo, Renew. Sust. Energ. Rev **42**, 8 (2015)
2. F.C. Zeng, J.L. Xu, Therm. Sci **22**, 13 (2018)
3. D. Marvaniaa, S. Subudhi, Renew. Sust. Energ. Rev **70**, 12 (2017)
4. A. Papson, F. Creutzig, L. Schipper, Transport. Res. Rec **93**, 8 (2010)
5. Z.G. Xu, X.P. Xie, IEEE, 5 (2009)
6. Z. Zhang, R.B. Jia, Q.H. Yu, M.L. Cai, AMM, **278-280**, 5 (2013)
7. R.Z. Song, X. H. Fu, M.L. Cai, AMM, **278-280**, 8 (2013)
8. C.Y. Huang, C.K. Hu, C.J. Yu, C.K. Sung, Energies **6**, 15 (2013)

Proposal for a solar fruit dryer design with organoleptic properties recovery system

Asmat-Campos, D.[1,2] , *Angel* Carreño-Ortega[2]

[1]Private University of North (UPN), Trujillo, Perú
[2]Department of Engineering, Higher School of Engineering, Agri-Food Campus of International Excellence (CeiA3), University of Almería, Ctra. Sacramento s/n, 04120 Almería, Spain.

Abstract. With the passing of time, the demand for dehydrated organic products has increased, ranging from vegetables to fruits, however in the latter efforts to achieve a quality product have led to the use of industrialized systems which depend on the use of fossil fuels or systems that use electric current which greatly increase production costs. At present, various solar dryer models have emerged, having achieved an optimal performance in what corresponds to the processing of vegetables, however in the processes of dehydration of fruits the results have not been satisfactory, due to the loss of its color, taste, aroma and texture (organoleptic properties). Faced with this problem, an innovative design of solar fruit dehydrator is proposed, which has two objectives, the first related to recovering the organoleptic properties through a system of condenzation, cyclic dehydration and the synchronization of the calorific balance, and the second one focused on the improvement of thermal efficiency, through the use of some heat piles and a trombe wall. The heat transfer fluid was improved, adding silver nanoparticles (silver nanospheres), which were synthesized using the green chemistry method; the efficiency of the heat transfer fluid was increased by 30%.

1 Introduction

The drying process occurs naturally, an example of this is when the seeds are dried to reduce their mass and can be lighter for dispersion in the air. The key strategic objective in crop preservation is to reduce the number of undernourished people in the wider world [1]. Since past times, open-air drying has been practiced, where it consists only of exposing the harvested product to solar radiation, either by placing it on the ground or a mat, however, the product is prone to great losses, due to an uneven removal of moisture, and exposure to insects and rodents [2]. Dehydration is the most common form of food preservation, because it extends the life of food [3]. Water is the main component present in fruits, which is why it is essential to be able to control it in order to avoid organic and nutrient deterioration [4]. With a system of dehydration it is possible to preserve any type of organic material, this by means of the reduction of the levels of humidity, so that the microorganisms can not develop [5]. At present there are many artisanal ways of drying fruits using the sun, because it is considered a fairly cheap method [6].

In developing countries more than 80% of organic foods are produced by small farmers [7], however, they do not use solar drying or dehydration because they can not achieve a quality product, with several disadvantages in terms of of product competitiveness. There are several studies where they have shown that solar dryers can be effective [6].

The success in solar drying processes comes from transforming products such as vegetables which do not need much of a meticulous process in order to not lose their organoleptic properties. In the fruit export process, it happens that, depending on the refrigeration system, the cold line is sometimes not adequate, leading to losses that reportedly reach 21%, while other sources estimate a value between 40 and 50% [8].

The process of dehydration is of the dual type, since it consists of the transfer of heat to the product located in a dehydration chamber, and the massive removal of moisture from the product from the interior to the exterior (surface) of the environment [9]. As in drying, dehydration leads to greater preservation of the product, becoming a method that manages to eliminate a large percentage of moisture content, however, there are few systems that achieve a product that meets quality standards, even more in the case of fruits, because the processes of heat transfer are not taken into account gradually, which leads to the loss of organoleptic properties. For years, systems and processes have been innovating under the implementation of technologies that depend on fossil fuels and / or the use of conventional electric energy by boilers, which have increased production costs by a large percentage, making it difficult for the consumer to be able to acquire it.

There is a diversity of solar drying systems, which have been innovated through better heat transfer by convection and radiation, however, they have not been successful in relation to products with high fructose content, however, several researchers have developed the so-called solar dehydrators, which present a greater efficiency at thermal level, which became a much more viable option in comparison with high technology systems [10].

In this article, an innovative design of solar dehydrator is proposed, the same one that has two objectives, the first related to the preservation of the organoleptic properties in the process of dehydration of the fruit through the use of a system of recovery of aroma in the process of synchronized condensate and heat reflux, and secondly, the use of auxiliary backup systems to supply synchronized heat in periods of low intensity of solar radiation, in addition to increasing the heat transfer efficiency of the heat transfer fluid, by using of nanofluid based on silver nanoparticles, the same ones that were synthesized as part of this work, under a method denoted as green chemistry.

2 Methodology

In the agro-export market, the preservation of the product plays a fundamental role, having to use industrial refrigeration systems in order to achieve a product in good condition at the destination. Dehydration is an accepted alternative for large exporters, because a large range of fruits are accepted by the consumer in this presentation and in addition to modern freeze drying systems have achieved a product of excellent quality, however production costs they increase abysmally, that is why the proposal to use solar energy through an innovative system, would be efficient at a cost - benefit level, for this it is essential to establish the various variables that will intervene in the process, including relative humidity , temperature, volume of air and heat given.

Theoretically, it has been established that the synchronized increase of the heat transferred to the product leads to a gradual process of dehydration of the product, thus generating the non-rupture at the cellular level of the fruit, in turn the condensation system of the moisture of the product and the reflux of the product. heat transferred to complete the process, without suffering the sudden change of heat. In addition, the increase of heat through alternative heat sources to the system would generate an improvement in the efficiency of the equipment, however it is essential that the dehydration process be continuous (even in hours where there is no solar radiation), which is why it has been taken into account the adherence of a system called "Thermobank", which consists of some thermal batteries which store heat and transmit them by convection and radiation to achieve a thermal equilibrium in hours where the temperature tends to fall.

2.1 Dehydration equipment:

The innovated model, presents simultaneously systems which generate a heat transfer, which are controlled by temperature and humidity sensors, and a moisture condensation system, which guarantee a good dehydration process and therefore a good product quality. In figure 1, the design of the equipment is appreciated. In the figure it can be seen that the system is composed of heat sources such as a parabolic cylindrical collector (by heating a high density fluid) reserving that heat in a thermobank, and a "trombe" wall, installed in one of the the rear faces of the dehydration chamber. Regarding the system of recovery of aromas and organoleptic properties, it consists of a system of condensation and drainage of moisture, recovering directly by means of a closed circuit assisted by a centrifugal fan, the aroma of the dehydration chamber. In the latter, the circuit has a series of passage valves, which are synchronized according to the process. The design also covers the accumulation of thermal energy for hours where solar radiation is insufficient.

2.2 Process of operation:

The designed equipment is specific for the dehydration of fruits by means of a system of calorific contribution assisted by two sources (trombe wall and parabolic cylindrical collector), It should be noted that this proposal is part of a patent by the authors of this research paper (Patent N° P201731235), in turn a thermobank is adapted to achieve the calorific balance by reducing the high temperatures obtained in parabolic cylindrical collector (when the intensity of solar radiation is high) and / or as a supplement to the reserve of heat energy in hours where solar radiation is weak (using heat piles, stones).

Inside the dehydration chamber there is a duct with an outlet in the upper part of the dehydration chamber and an entrance in the lower part of the chamber. It consists of three valves which act synchronously to let the moisture of the product pass, condensing it, draining it and finally return it to the system, all of it forced by a centrifugal fan, thus avoiding the loss of the aroma of the product (fruit) and other organoleptic properties; the system depends on two fundamental variables: temperature and humidity, for this case they will be

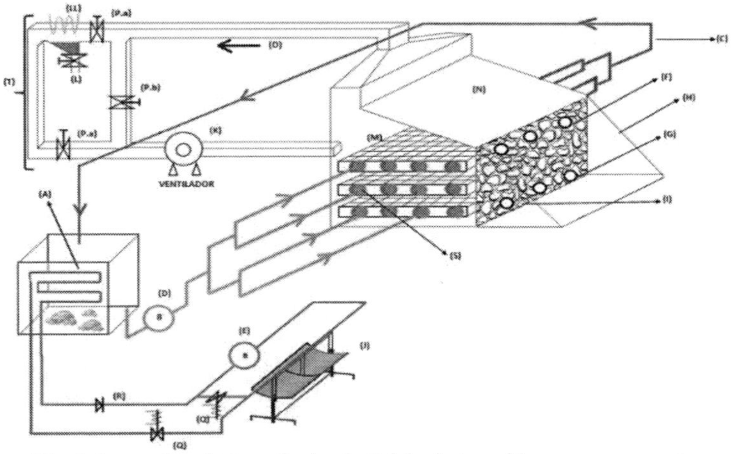

Fig. 1. Innovative design of solar fruit dehydrator with aroma recuperator.

monitored by humidity and temperature sensors with data logger, provided with electric current by photovoltaic panels.

The calorific contribution that has the dehydration chamber is given by two coupled systems, the first: The trombe wall (wall made of stones joined with concrete, with holes in the top and bottom, with the front side painted matt black, At the same time, on the outside, a fully waterproofed glass house to achieve a greenhouse effect and save heat in the thermal stone piles) (Figure 2), this system provides hot air by natural convection to the dehydration chamber (with greater intensity in hours where there is direct solar radiation and with less intensity where there is no solar radiation -night). Regarding the second source of heat input: The parabolic cylindrical collector is provided with the passage of an oil (heat transfer fluid) through the focal axis (which will act with a whole sequence of solenoid and check valves); however, the temperature is balanced by adapting a thermobank, which contains some stones (thermal batteries), passing to heat a second heat transfer fluid (water), being transported by copper pipes (properly thermally insulated) for conduction to the dehydration chamber, which provides the product with heat by thermal radiation; the fluid once transferred its calorific contribution to the product, will return to the thermobank (by natural convection) for its subsequent heating, working by means of a closed circuit.

The dehydration chamber has metal mesh trays placed one on top of the other, and in the lower part passes copper pipe where the second heat transfer fluid flows, thus providing an ideal temperature; At the same time, the calorific contribution of the trombe wall described above is added, thus improving the production speed of the dehydrated fruit.

The moisture emanating from the fruits as an effect of the dehydration process is conducted by the duct described above gradually, this process will be by forced draft, by coupling an air pump.

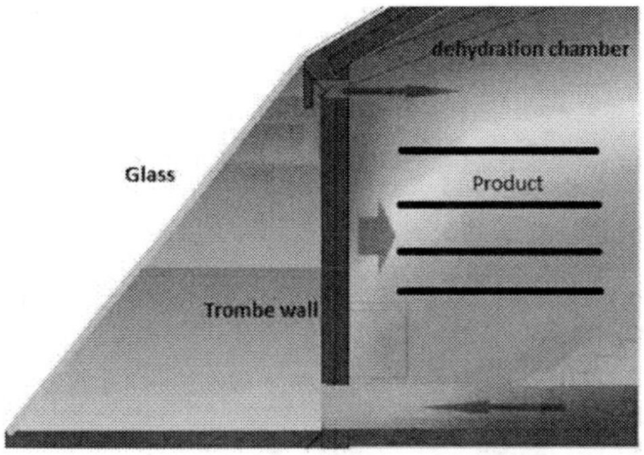

Fig. 2. Heat transfer through the trombe wall

This system solves the typical technical problem of solar dehydrators, such as inefficiency with respect to production time, counting with two sources of heat input, being able to work in high, medium and low intensity solar radiation conditions; However, it also highlights what relates to the recovery of aroma, because it is an important feature of arduous demand in export processes of dehydrated products or flour.

2.3 Explanation of the innovative solar dehydration equipment

Solar dehydrator fruit equipment, with aroma recovery system. (Figure 3).

The design consists of two structural complexes, which in turn raise two objectives, the first to recover the organoleptic properties of the fruit (specifically the aroma), and the second is to provide an efficient calorific contribution from solar energy, this last able to provide heat energy even at daytime where it is cloudy and / or at night.

https://doi.org/10.1051/e3sconf/20198002003

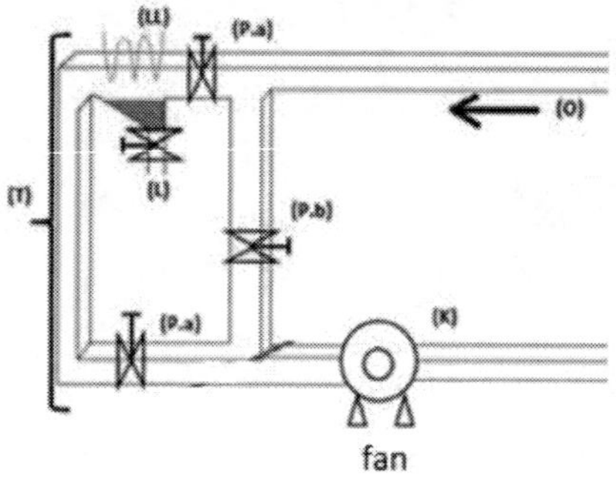

Fig. 3. Design of the organoleptic properties recovery system.

For the system of aroma recuperator (T), it will be able to capture the humidity coming from the fruit, coming from a first outlet pipe (O), going through some valves (Pa) Y (Pb), which will synchronize the passage of the humidity, until arriving at the condensation process by assisted conventional refrigeration (LL), and drainage (L); the process will be forced by centrifugal fan (K). Secondly, with regard to the efficient calorific contribution, a design of thermal bank of heat or "thermobank" (A) has been incorporated in a traditional way, which allows under a conduction heat transfer process (heat emitted by pipeline). (E) coming from the parabolic cylindrical collector (J) to the water of the thermobank (A) for its later conduction through the pipe (D) to the dehydration chamber allowing to "calibrate" the ideal temperature and in turn save the heat energy in some "Heat batteries" (stones) located in the thermobank, the latter with the objective that the dehydrator work in hours where there is no intensity of solar radiation (night) and / or this cloudy (low intensity of solar radiation), and emit its heat energy due to the effect of thermal equilibrium (zero law of thermodynamics) with the same destination (dehydration chamber), thus achieving the presence of controlled heat in the product that is e n a constant dehydration process.

2.4 Description of the drawings:

Figure 1. Solar dehydrating fruit equipment with aroma recovery system.

(A) Thermobank, (T) Scent recovery system, (C) Duct outlet (cold water) from the dehydrator, (D) Duct entrance with hot water to the dehydrator, (E) Piping with heat transfer fluid (oil) from the parabolic solar collector, (F) Hot air inlet holes, (G) Cold air outlet holes, (H) Glass hut with tilt angle, (I) Wall of stones joined with concrete, (J) Parabolic cylindrical collector

(K) Centrifugal fan, (L) Humidity drainage, (LL) Humidity condenser per conventional refrigeration cycle, (M) Metal mesh trays, (N) Solar dehydration chamber, (O) Main pipeline, (P.a), (P.b) Valves, (Q) Solenoid valves, (R) Valve check, (S) Copper piping (hot water passage).

The modeling of the innovative design was made through the analysis of a native fruit of Peru called lúcuma (*Pouteria lúcuma*), with the aim of giving it an added value, due to the great international demand. For this case, the calculations related to the useful solar radiation were made by means of the equation (1) foreseen for the city of Trujillo, place that owns part of the greater production of this fruit, the city in study has a great agroexporter potential, being its favorable weather conditions (minimum) to be able to bet on applying the system envisaged. Trujillo (department of La Libertad) is located 34 msnm, in the north coast of Peru, taking into account that the average solar radiation intensity is 550 W/m^2, and its temperature varies between 18 °C and 30 °C in winter and summer respectively. It should be noted that in Peru there are areas in the sierra area, whose intensity of solar radiation reaches up to 1700 W/ m^2, additional motivation that was taken with the aim of achieving the same thermal conditions to achieve dehydration of products with total normality.

$$S = (I_t)(k)(\tau)(\alpha_b) \qquad (1)$$

The results showed that for the study area the useful solar radiation with which it would be counted (on average per year) is 390.45 W/m^2, in addition to this there is the calculation related to the useful power, both for the wall system trombe the heating of an air volume of 13.53 m^3, and a concrete trombe wall with masses of stones, achieving a value of 202.22 W/h and 420 W/h. respectively; The following equations are those used for the respective calculations of the useful power of the trombe wall and the parabolic cylindrical collector (CCP) respectively.

$$Q_u = (A_c)(F_R)[(S) - U_L(T°_{fi} - T°_a)] \qquad (2)$$

$$q = \frac{T - T_\infty}{\frac{\ln r_3/r_2}{(2)(\pi)(l)(K_v)} + \frac{1}{(h)(2)(\pi)(l)(r_3)(180/360)}} \qquad (3)$$

As part of the mathematical analytical protocol of the calculations, it was also taken into account, for the case of the CCP, the calculation of the geometric concentration ratio, the solar energy incident on the collector and the amount of useful thermal energy supplied by the equipment in mention.

2.5 Improvement of the efficiency of the heat transfer fluid, used nanoparticles

The dehydration system will have as energy sources both the trombe wall and the parabolic cylindrical collector, for the latter it is necessary to have a heat transfer flow, that is why the oil is used, however it is planned to add silver nanoparticles to the oil (silver nanospheres) the same ones that were chemically synthesized using silver nitrate ($AgNO_3$) as a precursor agent, and reduced using natural extract of the blueberry fruit, which makes it a green and non-polluting synthesis process. The results have been quite favorable when using this organic agent as a reducer, achieving nanospheres with a diameter that are within the range of 15 to 60 nm. Characterizations were made to the nanoparticles by X-ray diffraction, showing the peaks belonging to the crystalline structure of the silver, dynamic light scattering (DLS), which showed the diameters of the nanospheres, the latter corroborated by TEM images, to its Once spectrophotometry was performed, resulting in an optical absorbance peak of 420 nm (Figure 4), typical of spherical silver nanostructures.

The adhesion of this type of material to the heat transfer fluid improves a thermal efficiency of 30%.

The system of dehydration in general, consists of two thermal sources, in addition the adhesion of a thermobank (thermal bank), adds an efficiency in the productive process, increased the hours in which the product will be exposed to heat and carrying out the process dehydrated, the synchronization of the heat transfer is the key to carry out a moderate process of evaporation of the humidity of the fruit, adding a process of condensation of moisture and drained of it under processes already planned, stage which will have as objective not to lose the organoleptic properties of the fruit; the solar panels will be a voltage source for thermometric and humidity sensors.

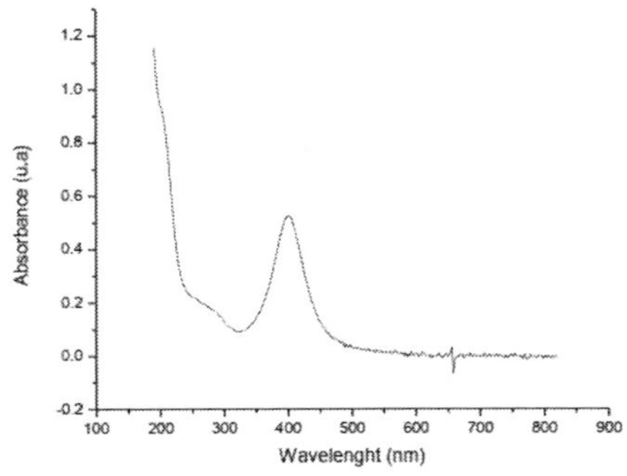

Fig. 4. Optical absorbance graph of the organic silver nanoparticles.

3 Conclusions

The innovative model of solar dehydrator, was able to form a hybrid system (cylindrical parabolic collector, wall trombe and thermobank), which add an energy of 622.22 w/h., and with daily dehydration capacity of 4.38 kg. of raw mass and 1.74 kg. of flour per day, being able to increase its capacity according to the area of the dehydration chamber; it would also reduce production costs by 23% compared to a conventional electric dehydrator; the benefit would also be reflected in the environmental field, reducing a total of 444.27 kg. of annual co2 with only one team, which is why the project could benefit from carbon credits. The use of silver nanofluids, increased the thermal conductivity capacity by 30%.

4 References

1. Anon., 2004. The State of Food Insecurity in the World, Food and Agriculture Organization of the United Nations Viale delleTterme di Caracalla, 00100 Rome, Italy.
2. Tomar, V., et al. Solar dryers for tropical food preservation: Thermophysics of crops, systems and components. Sol. Energy (2017).
3. Sarasavadia, P.N., Sawhney, R.L., Pangavhane, D.R., Singh, S.P., 1999. Drying behaviour of brined onion slices. Journal of Food Engineering 40, 219–226.
4. Saguy, I., Karel, M., 1980. Modelling of quality deterioration during food processing and storage. Food Technology 78, 84.
5. Mahmutoglu, Teslime, Ferhunde, Emir, Birol Saygi, Y., 1996. Sun/solar drying of differently treated grapes and storage stability of dried grapes. Journal of Food Engineering 29 (3–4), 289–300.
6. Pangavhane, D.R., Sawhney, R.L., 2002. Review of research and development work on solar dryers for grape drying. Energy Conversion and Management 43, 45–61.
7. Murthy, M.V.R., 2009. A review of new technologies, models and experimental investigations of solar driers. Renewable and Sustainable Energy Reviews 13 (4), 835–844.
8. Sodha, M.S., Bansal, N.K., Kumar, Ashvini, Bansal, P.K., Malik, M.A.S., 1986. Solar Crop Drying. CRC Press, New York.
9. Ekechukwu, O.V., Norton, B., 1999. Review of solar-energy drying systems II: an overview of solar drying technology. Energy Conversion and Management 40, 615–655.
10. A. Sharma, C.R. Chen y Nguyen Vu Lan. "Solar energy drying systems: A review". Renewable and sustainable energy reviews, vol 13 (2009), pp.1185 – 1210.
11. O. V. Ekechukwu. "Review of solar-energy drying systems III: low temperature air-heating solar collectors for crop drying applications". Energy Conversion and Management, vol 40, No 6. (1999), pp.657-667.
12. O. V. Ekechukwu. "Review of solar-energy drying systems I: an overview of drying principles and theory". Energy Conversion & Management, vol 40, No 6. (1999), pp. 593 613.
13. Ali Motaveli, Saeid Minaei y Mohammad Hadi Khoshtagaza. "Evaluation of energy consumption in different drying methods". Energy conversion and management, vol 52, (2011), pp. 1192 - 1199.
14. F. Rodríguez. Ingeniería de la industria alimentaria (III). Conservación de alimentos por reducción de su actividad de agua. Secado. Capítulo 5, 2002.
15. I. Doymaz. "Convective drying kinetics of strawberry". Journal Chemical engineering and processing, vol 47 (2008), pp. 914–91

E3S Web of Conferences **80**, 02004 (2019)
REEE 2018 https://doi.org/10.1051/e3sconf/20198002004

Study on architecture design of electroactive sites on Vanadium Redox Flow Battery (V-RFB)

Suhailah Sujali[1], *Mohd Rusllim* Mohamed[1], *Ahmed Nurye* Oumer[2], *Azizan* Ahmad[3] and *Puiki* Leung[4]

[1]Faculty of Electrical and Electronics Engineering, Universiti Malaysia Pahang, 26600 Pekan, Pahang, MYS
[2]Faculty of Mechanical Engineering, Universiti Malaysia Pahang, 26600 Pekan, Pahang, MYS
[3]School of Chemical Sciences and Food Technology, Universiti Kebangsaan Malaysia 43600 Bangi, Selangor, MYS
[4]Department of Materials, University of Oxford, Oxford, OX1 3PH, UK

Abstract. Numerous researches have been conducted to look for better design of cell architecture of redox flow battery. This effort is to improve the performance of the battery with respect to further improves of mass transport and flow distribution of electroactive electrolytes within the cell. This paper evaluates pressure drop and flow distribution of the electroactive electrolyte in three different electrode configurations of vanadium redox flow battery (V-RFB) cell, namely square-, rhombus- and circular-cell designs. The fluid flow of the above-mentioned three electrode design configurations are evaluated under three different cases i.e. no flow (plain) field, parallel flow field and serpentine flow field using numerically designed three-dimensional model in Computational Fluid Dynamics (CFD) software. The cell exhibits different characteristics under different cases, which the circular cell design shows promising results for test-rig development with low pressure drop and better flow distribution of electroactive electrolytes within the cell. Suggestion for further work is highlighted.

1 Introduction

Energy storage is recognised as an important technology to work with variable and intermittent in nature of renewable power sources. Therefore, energy storage has been introduced to retain and sustain the required operational and system reliability of the power sources [1][2]

Through various energy storage technologies, vanadium redox flow battery (V-RFB) shown in Fig.1[3] offers a promise because of its unique features by providing effective and simple operation, capability of decoupling high power and energy, may operate at room temperature, fast response and recharging contenders. Furthermore it comprises an excellent chemical stability that shows an extremely long-round-trip cycle, have long discharge times, good in modular design, exhibits for highly reversible redox kinetics, enable the technology for large-scale applications with theoretically reasonable and controlled maintenance cost compared to conventional battery[1][4–7]

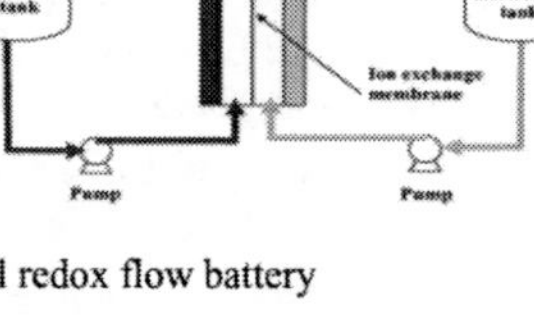

Fig. 1. Unit cell redox flow battery

As the name applies, flow battery is an electrochemical cell that produces electrical output through electrochemical reactions of two electrolytes that flow within the cell. Most of the researches in the past focused on improving poor kinetic reactions [8], conductivity [9], redox couples [10] electrodes [11] and battery characterisation [12]. A more recent works by Houser [13] *et al.* and Kumar [14] *et al.* were focusing on improving mass transport issues in redox flow battery. Mass flow is play an important role in battery to perform well [15][16]. The basic idea of study is to overcome the mass transport polarization issues and outcomes by introducing new design channel that may distribute the electrolyte well with reasonable pumping energy required. Recently, researchers systematically discuss about the electrode porous (electroactive sites) that acts as key in V-RFB systems[11][17][18].

In the conventional flow field of porous electrode, the electrolyte is directly pumped through the electroactive sites electrodes. This may results in unevenly distributed flow within the cell because the electrolyte flow through random pores of the electroactive sites part. This may causes pressure drop arising together with flow direction and this lead to a mass transfer losses and deterioration problems when the electrolyte flow into the electrode just from one lateral side. Previous researchers [19][20] have explored on creating new and appropriate flow channel for control the electrolyte flow and directly gives a positive feedback on improvement of V-RFB performance. There are literatures on several experimental and numerical investigation of the V-RFB design system. Xu *et al.* [19][21] proposed experimentally and numerical investigation on architecture of redox flow battery. The study focused on evaluation of vanadium redox flow battery performance based on flow of electrolytes within the cell, focusing on with and without flow field electrode configurations supported by experimental data and numerically designed three-dimensional model of V-RFB. The study concluded that energy efficiency of V-RFB is much higher by 5% when electroactive sites of flow field was applied as compared to energy efficiency without flow field.

Conversely, Wang *et al.* [22] studied on parallel flow design configurations concluded that energy efficiency of V-RFB could be further be enhanced by 5% in staggered arrangement as compared to symmetrical arrangement. Wang *et al.* claimed that design of flow field for V-RFBs is an important areas that needs more attention for further investigation and it was concluded that limited work in this area of research has been conducted. Fig.2 depicts different electrode configurations that have been used in previous studies, namely no flow flied (plain)-, parallel- and serpentine-electrode configurations.

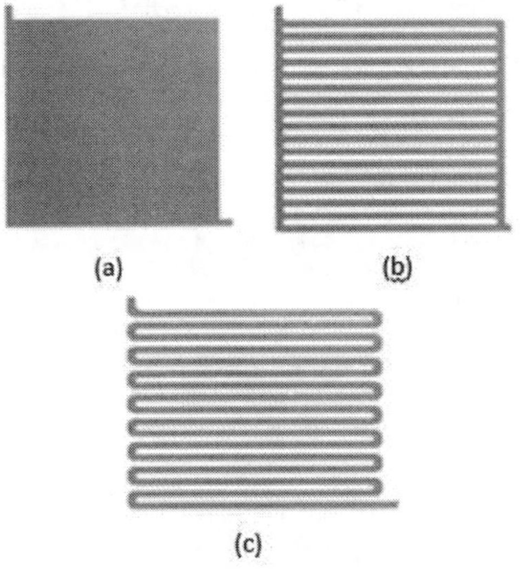

Fig. 2. Schematic drawing of three type of flow field (a) no flow flied- (b) parallel- (c) serpentine- configurations

On the other hand, Bortolin *et al.* [23] and Khor *et al.* [8] introduced a methodology included several performance indicators by using computational fluid dynamic simulation on determining the flow distribution in a particular geometry of V-RFB cell. Ressel *et al.* [24] studied the influence of different flow cell geometry design by introduced a tubular cell design with a flow by electrode configuration on the performance and system efficiency of V-RFB cell.

Based on the brief review, it is important to note that both flow field electrode configurations and architecture design of the cell play important roles for future advancement of flow battery technology.

Nevertheless, apart from Khor *et al.* which worked on rhombus shape electrode geometry design, all other highlighted studies were very much focused on square-shape based cell geometry design.

This study attempts to extend the works in both flow field electrode configurations and cell design geometry; looking into perspective of effect of flow distribution within three different cell design i.e. square shape, rhombus shape and circular shape electrode geometry design. The first part of studies were focused on identifying the best shape to be implemented in V-RFB cell design with respect to flow distribution within the cell based on no flow field electrode configurations. Extended works were focused on application of the best shape of electrode geometry design in further two electrode configurations to identify the lowest pressure drop which could resulted in better performance of the V-RFB.

2 Development of V-RFB cell in Computational Fluid Dynamics (CFD)

2.1. Simulation details

Numerical simulations were carried out and solved using the ANSYS 16.2 package with a combination of the self-written source terms in console zone. Three dimensional models were built in fluid flow (fluent) under analysis systems for handling the flow in the porous carbon electrode. The fluid model used is sulfuric acid solution diluted with vanadium oxides. The fluid was considered to be incompressible and isothermal, with density and dynamic viscosity were 1334.9 Kgm^{-3} and 0.0024 Pa respectively.

The model domains were developed based on a 100 cm^2 (10 cm x 10 cm) active sites of electrode with a thickness of 3 mm with 0.00424 m cross section area of the channel. The details operating parameter in geometry and setup parameter are shown in Table 1. This dynamic model were simulated with different velocity inlet condition that has been calculated analytically and a specified uniform flow velocity at the inlet and 0 Pa pressure at the outlet was set in the boundary condition. No slip condition on the walls and no flux condition were applied to the domain surfaces exceptional for inlet outlet flow channels. In all simulations, the fluent was set with double precision and pressure-based solver. The porous media model was evaluated with a pressure-velocity coupling algorithm SIMPLE (Semi Implicit Method for Pressure-Linked Equation) in solution methods. The CFD simulation were conducted at different flow rates, flow with/without distribution channel and with different electrode compartment design in order to evaluated the best configuration for transport phenomena and performance in the V-RFB.

Table 1 Operating parameters of CFD simulation of 100 cm^2 V-RFB cell

Parameters	Symbols	Value	Unit	Ref
Electrode thickness	δ_e	3	mm	

Electrode height	h	100	mm	
Electrode width	w	100	mm	
Polymer electrolyte membrane	δ_{mem}	3	mm	
Operating temperature	T	288	K	
Electrode porosity	ε_0	0.7	ι	[21]
Electrode fiber diameter	d_f	17.6	µm	
Electrolyte volumetric flowrate	Q	5, 10,15, 20, 25,30	cm^3s^{-1}	
Electrolyte density	ρ	1334.9	Kg m^{-3}	
Electrolyte kinematic viscosity	v	$3.2x10^{-6}$	m^2s^{-1}	
Carman-kozeny constant	K_{CK}	5.55		[21]
Carbon density	ρ	2240	Kg m^{-3}	
Specific heat of carbon	$c\rho$	710	$Jkg^{-1}K^{-1}$	
Thermal conductivity of carbon		0.2	$Wm^{-1}K^{-1}$	
Vanadium concentration	c	1.6	mol m^{-3}	
Electrolyte dynamic viscosity	μ	0.0024	Kg m^{-s}	
Thermal conductivity of electrolyte		0.67	$Wm^{-1}K^{-1}$	
Reservoir volume	V_C	200	ml	
Rib width		3	mm	
No of channels/bends		16		

2.2 Pressure drops in cell

Pressure drop is a crucial issue in the system for determining the smooth continuous flow circulation of the liquids inside the cell stack. At variance of flow pattern design in the cell electrode (active area) may give a different flow distribution and pressure drop. In this study, two conditions were set; at lower flow rate, the flow was consider laminar flow whereas higher flow rate was considered as turbulent flow. This consideration of flow was supported by the calculation by referring Reynolds number formulae.

The Reynolds number were determined respectively are shown in Equation 1. In this paper, various flow pattern design were considered including without distribution channel (direct feeding) in a porous media electrode and indirect feeding channel; a parallel and serpentine pattern.

$$Re_{efb} = \frac{\rho_{efb} VD}{\mu_{efb}} \quad (1)$$

Pressure drop result is shown for a pumping energy required in a redox flow battery cell. The lesser pressure drop of the cell battery, the lower consumption of pumping energy needed. At variance of flow pattern design in the cell electrode (active area) may give a different flow distribution and pressure drop. Differential pressure (Δρ) is a pressure between inlet and outlet (over the length of the electrode, L), and this result comes from the relationship of the viscous resistance flow, permeability of the electrode material and viscosity of the electrolyte. A pressure drop through this electrode can be resolved according to the Darcy's Law by including a flow rate (Q) and viscosity of the electrolyte (η_e) also as shown in Equation 2:

$$\Delta P = \frac{\eta_e}{K_e(\varepsilon)} Q . \frac{L}{A} \quad (2)$$

The permeability (K_e) is dependent on the porosity of the electroactive sites of carbon electrode. Flow cross-sectional area (A) changes with different cell geometry (active sites).

2.3 Computational domains and boundary conditions

A symmetry boundary condition were elucidated to set the operational point of V-RFB cell that consist of inlet velocity magnitude, gauge-pressure at outlet region, symmetry planes and wall condition surface .The no-slip boundary condition was applied on the electroactive sites of the electrode excluded the inlet and outlet regions. Next, on the solution method setup part, pressure-velocity coupling was selected and performed by the SIMPLE algorithm and the second-order upwind scheme was used in a computational domain. Finally, numerical analysis were simulated through continues running by maintaining a steady state conditions of the operation of the battery cell and mass-weighted values applied also monitored.

3 Result and Discussions

With the aim to investigate the influence of the cell's electrode geometry design on flow distribution within the cell, flow rates at 5 $cm^3 s^{-1}$ to 30 $cm^3 s^{-1}$ were fed into three proposed 100 cm^2 square shape, rhombus shape and circular shape electrode geometry design. Fig. 3 indicates the shapes plays significant parts affecting the flow distribution within the three proposed electrode geometry designs when applied through no flow field (plain) electrode configuration at flow rate of 30 $cm^3 s^{-1}$. The result shows that circular shape design gives the best uniformity electrolyte distribution as compared to square shape and rhombus shape electrode geometry design. Other results for different flow rates will be published in forthcoming extended paper.

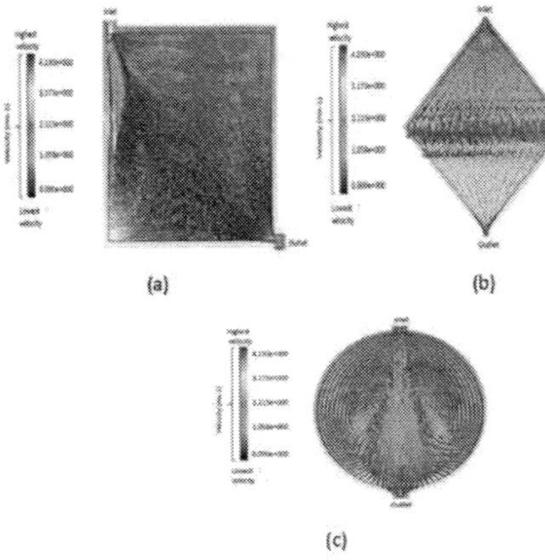

Fig. 3 Velocity vector shows flow distribution for 100 cm^2 three different geometrical cell design at same 30 cm^3 s^{-1} flow rates applied (a) square cell (b) rhombus cell (c) Circular cell

Moreover, pumping power is one of the impact parameter for determining a better RFB system. It should be noted that, lower pressure drop indicates a low pumping consumption. In this study, pressure drop has been calculated to determine which proposed geometrical cell design consume a lesser pumping power. In this case, area-weighted average of static pressure has been computed for each geometrical designs. Fig. 4 shows the pressure drop for each geometrical designs at various flow rate of 5 cm^3 s^{-1} until 30 cm^3 s^{-1}. It should be noted that, the flow rate increases in accordance to applied flow rate, indicating of uniform distribution of electrolytes within the cell with the expense of higher pumping losses; higher pumping power required is proportional to higher pumping losses which could affecting in overall efficiency of V-RFB system. The results shows that the square flow frame has the highest pressure drop (17.3 KPa) at flow rates 30 cm^3s^{-1}, followed by rhombus cell (16.5 KPa) and circular shape is (12.2 KPa).

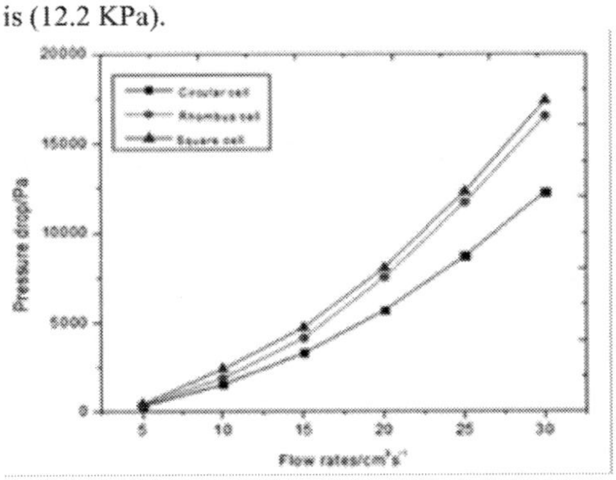

Fig. 4. Comparison of pressure drop within the three different 100 cm^2 geometrical cell designs i.e. square shape, rhombus shape and circular shape cell geometry design at various flow rate of 5 cm^3 s^{-1} until 30 cm^3 s^{-1}

Furthermore, with circular shape cell design represent best result for both flow distribution and pressure drop at various flow rates, the study is extended to determine which flow field configurations would be the best to be applied on 100 cm^2 circular shape cell design.

Fig.5 represents the pressure drops for no flow field porous medium model and two different flow pattern at the flow rates 5 cm^3 s^{-1} until 30 cm^3 s^{-1}. The result of the numerical simulation indicates that the parallel flow field shows the lowest pressure drop at 253 Pa whereas the no flow field and the serpentine distribution channel has represented a 422 Pa and 2469 Pa pressure drop. This indicates that serpentine flow channel has a highest pressure drops and no flow field design is two times higher than parallel design. By comparing the parallel and no flow field designs, it should be noted that at lowest flow rates, by adding flow field at bipolar plates it can significantly reduce a pressure drop. Nevertheless, when the flow rate is increased, for example of 30 cm^3 s^{-1}, the pressure drop for the porous electrode and two types of flow patterns increase accordingly as shown in Fig.5. For the first two fields, the pressure drop increases from 253 to 12432 Pa and from 2469 Pa to 45701 Pa respectively whereas without flow field electrode it increases from 422 Pa to 17375 Pa. Details data of effect on pressure drops for respective square shape, rhombus shape and circular shape cell geometry design with respect to flow rates 5 cm^3 s^{-1} until 30 cm^3 s^{-1} will be discussed in forthcoming paper.

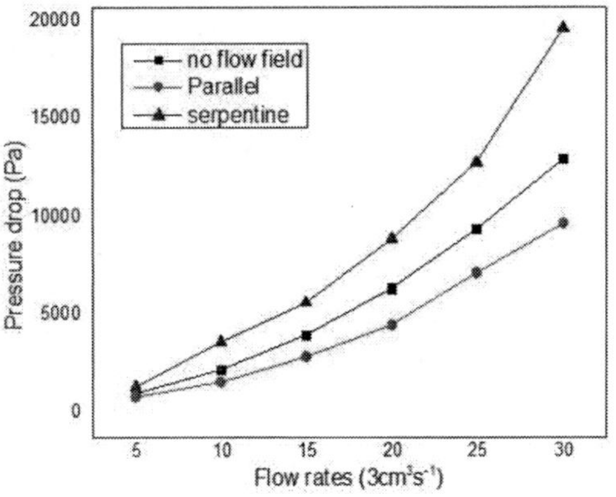

Fig. 5. Computationally pressure drop data for no flow field, parallel- (c) serpentine-configurations for 100 cm^2 circular shape cell geometry design with respect to flow rates 5 cm^3 s^{-1} until 30 cm^3 s^{-1}

4 Conclusions

This study demonstrates that the circular shape cell geometry design has most uniform electrolyte distribution within the cell at applied 30 cm^3 s^{-1} flow rates as well as lowest average pressure drop as compared to square and rhombus cell design counterparts, whereas parallel flow field configurations represents the best options for 100 cm^2 circular shape cell geometry design of V-RFB at 5 cm^3 s^{-1} - 30 cm^3 s^{-1} flow rates. Further works on effect of cell architectures to the performance of V-RFB with respect to coulombic, voltage and energy efficiency will be discussed in forthcoming paper.

Acknowledgements

This work was supported by Malaysia's Ministry of Education (MOE) - Fundamental Research Grant Scheme (FRGS) - RDU 150123, Research Acculturation Collaborative Effort (RACE) - RDU151316 and Postgraduate Research Scheme - PGRS170352, Universiti Malaysia Pahang.

References

1. M. Guarnieri, P. Mattavelli, G. Petrone, and G. Spagnuolo, "Vanadium Redox Flow Batteries," no. december 2016, pp. 20–31, 1932.
2. T. M. I. Mahlia, T. J. Saktisahdan, A. Jannifar, M. H. Hasan, and H. S. C. Matseelar, "A review of available methods and development on energy storage; Technology update," *Renew. Sustain. Energy Rev.*, vol. 33, pp. 532–545, 2014.
3. C. P. De Le and F. C. Walsh, "Redox flow cells for energy conversion," vol. 160, pp. 716–732, 2006.
4. M. Rusllim Mohammad, S. M. Sharkh, and F. C. Walsh, "Redox flow batteries for hybrid electric vehicles: progress and challenges," *2009 IEEE Veh. Power Propuls. Conf.*, no. August 2016, pp. 551–557, 2010.
5. A. Z. Weber, M. M. Mench, J. P. Meyers, P. N. Ross, J. T. Gostick, and Q. Liu, "Redox flow batteries: A review," *J. Appl. Electrochem.*, vol. 41, no. 10, pp. 1137–1164, 2011.
6. G. R. Fisher, M. Ieee, M. R. Anstey, V. V Viswanathan, and M. L. Perry, "Redox Flow Batteries : An Engineering Perspective," pp. 1–24, 2014.
7. T. Jyothi Latha and S. Jayanti, "Hydrodynamic analysis of flow fields for redox flow battery applications Batteries," *J. Appl. Electrochem.*, vol. 44, no. 9, pp. 995–1006, 2014.
8. A. C. Khor *et al.*, "Numerical investigation on serpentine flow field and rhombus electrolyte compartment of vanadium redox flow battery (V-RFB)," *ARPN J. Eng. Appl. Sci.*, vol. 11, no. 10, 2016.
9. T. Mohammadi, "Characterisation of novel composite membrane for redox flow battery applications," vol. 98, pp. 77–87, 1995.
10. P. K. Leung, M. R. Mohamed, A. A. Shah, Q. Xu, and M. B. Conde-duran, "A mixed acid based vanadium e cerium redox fl ow battery with a zero-gap serpentine architecture," vol. 274, pp. 651–658, 2015.
11. P. Leung, J. Palma, E. Garcia-quismondo, L. Sanz, M. R. Mohamed, and M. Anderson, "Evaluation of electrode materials for all-copper hybrid fl ow batteries," *J. Power Sources*, vol. 310, pp. 1–11, 2016.
12. M. R. Mohamed, P. K. Leung, and M. H. Sulaiman, "Performance characterization of a vanadium redox flow battery at different operating parameters under a standardized test-bed system," *Appl. Energy*, vol. 137, pp. 402–412, 2015.
13. J. Houser, A. Pezeshki, J. T. Clement, D. Aaron, and M. M. Mench, "Architecture for improved mass transport and system performance in redox fl ow batteries," *J. Power Sources*, vol. 351, pp. 96–105, 2017.
14. S. Kumar and S. Jayanti, "Effect of flow field on the performance of an all-vanadium redox flow battery," *J. Power Sources*, vol. 307, pp. 782–787, 2016.
15. X. Ma, H. Zhang, C. Sun, Y. Zou, and T. Zhang, "An optimal strategy of electrolyte flow rate for vanadium redox flow battery," *J. Power Sources*, vol. 203, pp. 153–158, 2012.
16. A. Tang, J. Bao, and M. Skyllas-Kazacos, "Studies on pressure losses and flow rate optimization in vanadium redox flow battery," *J. Power Sources*, vol. 248, pp. 154–162, 2014.
17. a C. Khor, M. R. Mohamed, M. H. Sulaiman, and M. R. Daud, "Packaging Improvement for Unit Cell Vanadium Redox Flow Battery (V-RFB)," no. 6, pp. 808–811, 2014.
18. C. Blanc, S. Member, and I. A. Rufer, "Multiphysics and Energetic Modeling of a Vanadium Redox Flow Battery," 2008.
19. Q. Xu, T. S. Zhao, and C. Zhang, "Performance of a vanadium redox flow battery with and without flow fields," *Electrochim. Acta*, vol. 142, pp. 61–67, 2014.
20. D. H. Jeon, S. Greenway, S. Ã. Shimpalee, and J. W. Van Zee, "The effect of serpentine flow-field designs on PEM fuel cell performance," vol. 33, pp. 1052–1066, 2008.
21. Q. Xu, T. S. Zhao, and P. K. Leung, "Numerical investigations of flow field designs for vanadium redox flow batteries," *Appl. Energy*, vol. 105, pp. 47–56, 2013.
22. Q. Wang, Z. G. Qu, Z. Y. Jiang, and W. W. Yang, "Numerical study on vanadium redox fl ow battery performance with non- uniformly compressed electrode and serpentine fl ow fi eld," vol. 220, no. September 2017, pp. 106–116, 2018.
23. S. Bortolin, P. Toninelli, D. Maggiolo, M. Guarnieri, and D. Del Col, "CFD study on electrolyte distribution in redox flow batteries," *J. Phys. Conf. Ser.*, vol. 655, no. November, p. 12049, 2015.
24. S. Ressel, A. Laube, S. Fischer, A. Chica, T. Flower, and T. Struckmann, "Performance of a vanadium redox fl ow battery with tubular cell design," *J. Power Sources*, vol. 355, pp. 199–205, 2017

 https://doi.org/10.1051/e3sconf/20198002005

Reliability framework for power network assessment

Sambeet Mishra[1], *Chiara* Bordin[2], *Jordi Mateo* Fornes[3], and *Ivo* Palu[1]

[1]Tallinn University of Technology, Estonia
[2]SINTEF Energy research
[3]Department of Computer Science, University of Lleida, Jaume II 69, 25001 Lleida, Spain.

Abstract. Reliability of power system in terms of investments in network maintenance and restructuring for power distribution network has gained importance due to increase in distributed generation. To determine the reliability of the power distribution network, the state of power apparatus, losses in the network and consumer satisfaction indices are key factors. Considering the aforementioned, this paper proposes a holistic reliability framework for power distribution networks. The framework lists the following factors: life cycle of power apparatus, environmental and sociological, node reliability, arc reliability. A case study for reliability evaluation is performed on a modified IEEE 14 bus network. Furthermore, multiple scenarios of generation fault or outage are studied and results are presented. The key contribution of this paper is to present a novel and holistic reliability framework to model distribution network.

1 Introduction

The aim of this paper is to develop an evaluation framework for reconfiguration of an electrical power system. Given study is influenced by the economical aspect of a network operator which implies to have the needed electrical, environmental, and economical indicators of an electrical power system to assess the optimal possibilities for reconstructing it. Many power system management utilities try to rationalize their network and optimize the total life-cycle costs of the components, due to the regulation of power quality and the reliability issues. For the given problem, many methods of assessment modeling have been introduced in the literature. System well-being method for power networks adequacy assessment using monte carlo simulation is presented, specifically for capacity reserve allocation [1]. This study expanded with renewable energy resources in [2]. This subject is further explored in the book series about reliable and sustainable power systems management [3]. Role of outage management and strategies are discussed here [4]. Reliability evaluation of power systems establishes measures to identify power interruptions and their implications [5]. However the approach is limited to power system events. For instance the state of the power apparatus is an important indicator to observe the system status. Detail concepts of power system reliability are presented in [6]. Reliability and connected parameters are introduced and and event tree analysis for event prediction is described here [7].

Many regionally owned utilities have been privatized, where the main concern are optimal investment planning for expansion and maintenance while maintaining the system reliability. Therefore, to improve and maintain power quality various regulation models have been introduced. These models enforce network companies to optimize their operations and schedule their maintenance activities without having to compromise the reliability nor the safety of the power system network. Reliability analysis is one of the key ways to inspect optimal asset management. A study has been conducted in [8–10] for evaluating the failure rates and reliability modeling for power distribution network. The reliability analysis has been utilized for optimal dis-connector allocation. In the analysis, the failure rates are constant for similar components, which are influenced by many different mechanical, environmental and electrical stresses. Usually for component failure models reliability calculations are based on the exponent distribution and failure rates are considered to be constant, which may be an inadequate approach. In some scenarios a scalar value is used for estimating the component failure rates while planning. Monte Carlo simulations are implemented to take into account effects of the surrounding, or enhanced component failure models which are based on constant component failure rates, to evaluate component related and the environmental aspects in reliability analysis. Another modeling approach is proportional hazard method, where age and other various additional information have been considered. Further information may include weather and environmental factors that affect the components. These models require significant data to be analyzed to investigate the essential dependencies affecting the component failure and malfunction. At the same time, Markov models are more commonly used, where the component failure rates are modeled by estimating the effects of the component faults for the system. Moreover, there are many factors that may lead to component failure such as weather condition and power surges. Therefore, predicting the component failure accurately is a challenge. As a consequence, the readiness of the power distribution network to optimally resolve the fault while ensuring uninterrupted power supply is the research objective.

1.1 Motivation for Reliability-based Network Decision-making

The main demand for reliability oriented network restructuring (RNR) is to have the estimates of failure rates which consider the main stress factors and the possibility to have first estimates from incomplete data to update values when more improved data is available. Components of distribution networks (DN) are modeled separately since the component failure rate is independent of factors that affect the power network. In this study, DN has been divided into four main components: overhead-lines (OHL), cables, transformers, and groups of switch-gear. For each component the main reasons for permanent faults and auto re-closings is determined. Separate failure rates for each component types are based on the reasons for a fault, e.g. aerial line overall failure rate is dependent on lightning, different weather conditions, and other fault related causes. The main stress factors that influence the failure rate, have been determined for all the reasons mentioned above. All the stress factors are classified into appropriate classes, for instance, the location could be by the seaside, near the forest or on a field. For all these classes a weight has been defined to represent the effects of a certain class to failure rate. For a total failure rate, temporary and permanent faults can be calculated. A practical approach on component modeling is to use the idea that parameters used in failure rate modeling should be possible to be affected by the selected planning strategies. Therefore, effects of the weather are included as an own parameter to failure rate modeling, however they are still included in condition information of the affect to the stress tolerance. Furthermore, the effects of component aging can be explained with other component related factors, therefore, life expectancy is included in condition weight information. Failure rate parameters must be determined before modeling methods can be used [11].

General failure rate of components were calculated as a weighted mean from failure rates of separate companies. Defined parameter groups are used to calculate the separate failure rates. Basic input data is the component information, i.e. component type, failure rate, and the network topology. Additionally, other information are needed regarding those factors that affect the analyses results, such as maintenance costs. In the enhanced radial reliability analysis, the network is with feeders and zones, which refer to a part of a feeder. In the given analysis, expected amount of permanent and temporary failures in a zone is calculated as an entirety of the individual network component failures. For a temporary fault, the whole feeder is experiencing the same short interruption. In the given analysis, experienced faults are defined for each load point. Determination of maintenance costs is done by analyzing the possible terrains where the components are located related to the total interruptions in the certain area. Therefore, the RNR framework can be expressed as an asset management model considering the Life-Cycle Assessment of power system components. To replace OHL's and underground cables, reliability of its reconfiguration is based on environmental and consumer preferences, N-1 criteria, and the objectives for minimizing the investment costs. Reconfiguration of networks is primarily done to accommodate new consumers, which is achieved by extending the existing node through a new arc, and to replace some existing out-of-date lines. Network operators can adjust the failure rate and reliability parameters with their own network data. Smart grid components can identify the fault region of the feeder and update it with secure supply of energy from the same power network. Reliability indicators mainly include measures of outage duration and its frequency, the amount of power or energy not supplied, and the number of customers involved in outages. These indicators are determined over predefined period of time, such as SAIFI, SAIDI, CAIDI, ENS.

The rest of the paper is organized as the following sections: reliability framework, power network weight evaluation. In reliability framework each element of the framework concerning the power network is presented. An example case of IEEE 14-bus network is presented and discussed. Subsequently we evaluate generation loss scenarios and the results are presented in the power network evaluation section.

2 Power System Reliability and Adequacy

Power system adequacy refers to the condition of a power network considering generation, transmission and distribution units. Power system reliability refers to the state of network to sustain flow of energy from point of generation to demand at any point in time. The relationship between reliability and investment cost is presented in fig. 1. The figure signifies with outages or otherwise power interruptions the reliability reduces and the investment increases. Therefore, the planner need to pay attention so as to optimally plan for the expansions or maintenance of the distribution network. The indicators to determine the condition of power system can be broadly classified in to five categories: life cycle of power apparatus, environmental and sociological, node reliability, arc reliability and node reliability concerning losses. The table 1 outlines the associated indicators. These parameters organize the indicators taking into account the network as a node-arc formulation.

2.1 Life cycle assessment (LCA)

Life cycle assessment (LCA) is a method to determine the environmental impacts from a product, a process, or an activity. It is also used to assess the remaining utilization life. Throughout the product lifetime the impacts mainly originate from the power losses during the use phase, although installation, maintenance, and dismantling also contribute

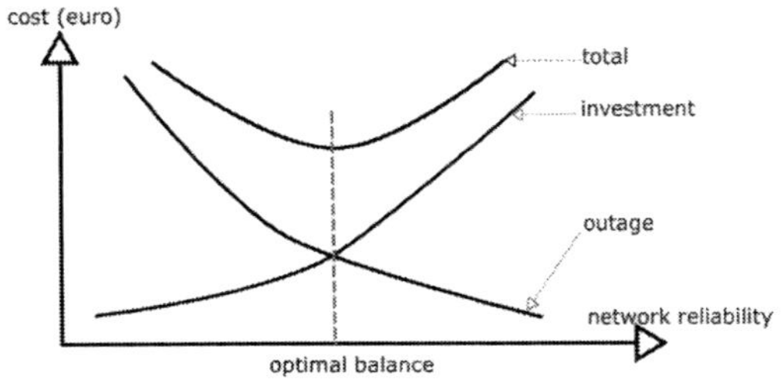

Figure 1: Relationship between reliability and cost

Table 1: Power system reliability and adequacy indicators

LCA	Environmental & Sociological	Node reliability indicators
Cable	Terrain	SAIFI value
OHL	Weather	SAIDI value
Transformer	Consumption	ENS value
Switchgear	Generation	Availability
Emissions		
Arc reliability indicators	**Node reliability indicators concerning losses**	
AIT	LOLE	
AIF	LOEE	
AID	EIR	

to it. Transmission and distribution assets have been comprised of power lines, cables, transformers, substations, and other electrical components to generate wide range of environmental impacts, such as the equipment emissions and material weight value. The life cycle stages viewed are interpreted as the production or the manufacturing phase of a product or its use phase. The used life cycle inventory consists of material requirement for grid components and their environmental impact. For all the components, the functional unit is one equipment operational during the lifetime.

Given network is a high and medium voltage network, with three distributed generation inputs. Biomass and oil shale produce the distributed generated energy, with an additional input from a submarine cable. The network nodes are depicted as substations with each bus-line having an ideal energy consumer as residential or commercial. The transmission line is either an underground cable or an overhead-line. The normalized weights are allocated to arcs and nodes based on the historical data and expert's opinion. For instance, an indicator depicting transmission line weight is valued at 4000.5 kg/km, although Switchgear emissions are valued at 185.38 kg CO_2 per transformer. Weights are normalized and translated to the cost of maintenance for the distribution system operator. In this method, dissimilar components can be compared based on the cost of investment.

2.1.1 Cable

Medium voltage power cables are characterized in [12]. In this study we concentrate on three different voltage levels, with each having one or two different types of cables, therefore five different cables are evaluated. Cables are chosen based on their suitable voltage level, cable diameter and the conductor and the insulation type.

A transmission line of 6 or 10 kV spanning over 9 km, a three-core cable is proposed with diameter of 3*70 mm^2. This cable weighs 5400 kg/km. Cable indicators in a network reconstruction evaluation are the conductor and insulation. In the example, conductor weight is calculated as the diameter of the conductor (i.e. Al conductor diameter = 9.8 mm) multiplied with conduction material density (Al density = 2.7 kg/km). i.e. Cable, which weighs 5400 kg/km has an aluminum conductor which weighs 124.7 kg/km and insulation of 5275.3 kg/km [13]. Different kinds of lines vary mainly in sizing of the cross section of the conductor, however the mentioned cross section is not the actual measurement of the wire, but rather the area of the conductor gives the needed cross section, such as aluminum, copper or steel. Data from cable producers is used to identify the correct amount of material input for each of the line types. For the power system network, at three different voltage levels, each voltage level is applied with power lines with the following conductors: aluminum (2.7 $\frac{gm}{cm^3}$), copper (8.89 $\frac{gm}{cm^3}$) and steel (7.83 $\frac{gm}{cm^3}$). The power line data is derived from manufacturer's data provided by ABB [14]. From the catalogue, the most suitable power lines are chosen, based on their voltage level and the conductor material. Considering the length of a given power line and the cross section provided by ABB, it is possible to find the weight of a conductor [14].

Given the total length of underground cables and overhead lines, and additionally it contains the calculated values of the power line conductor and insulation weight relative to their material. The insulation material composition is also derived from the manufacturer specifications [14]. Typically, the installation of cables requires extensive underground pathways. Due to insufficient data, these parameters are not included. However, some general construction processes are included, such as the weight difference and the relation to the actual line weight are applied. These electrical masts are calculated by their tension and sag related the length between each pole. Also, different height of masts is assumed for each voltage level. Poles are described by their suitable voltage level, material use, height, span, tension, and sag.

Table 2: Reliability assessment of power conductor

Voltage level (kV)	**Cable length (km)**	**Cable diameter (mm^2)**
6+10	9	3*70
Cable line weight (kg/km)	**Conductor diameter (mm)**	**Weight (kg/km)**
5400	9.8	646.5
Conductor material (Aluminum (Al))	**Insulation type**	**Weight (kg/km)**
Al	XLPE	5275.3

2.1.2 Transformer and switch-gear

Transformers used in this study are assumed to be ideal, without having to relate to the criterion N-1 (in case one

 https://doi.org/10.1051/e3sconf/20198002005

Node	1	2	3	4	5	6	7
Voltage level (kV)	110/10	110/35/10/6	110/35/6	110/35/6	110/10	35/6	110/35/6
Generation [0/1]	0	1	0	0	0	0	1
Generation, kWh	0	925	0	0	0	0	1000
Transformer [0,1,2]	1	2	1	1	1	1	1
Transformer type[1,2,3]	1	1+3	2	2	1	3	2
Life expectancy	0.2	0.2	0.2	0.2	0.2	0.2	0.2
Conductor weight (kg/Trfo)	0.61	1.00	0.57	0.57	0.61	0.00	0.57
Conductor weight (kg/kVA)	1.00	1.00	0.001	0.001	1.00	0.00	0.001
Trfo_oil weight (kg/Trfo)	0.56	1.00	0.22	0.22	0.56	0.00	0.22
Trfo_oil weight (kg/kVA)	1.00	1.00	0.00	0.00	1.00	0.0002	0.00
Energy demand (kWh)	250	275	275	325	300	350	350
Energy demand	0.00	0.25	0.25	0.75	0.50	1.00	1.00
SG type	1	4	2	2	1	3	2
Life expectancy	0	0	0	0	0	0	0
GWP	0.14	1.00	0.02	0.02	0.14	0.00	0.02
AP	0.22	1.00	0.03	0.03	0.22	0.00	0.03
NP	0.44	1.00	0.05	0.05	0.44	0.00	0.05
SF6 (% of all emissions)	0.00	1.00	0.11	0.11	0.00	0.02	0.11
Total value:	0.15	1.00	0.05	0.05	0.15	0.01	0.05
Terrain	4.00	9.00	5.00	7.00	4.00	5.00	7.00
Terrain coefficients	0.38	1.00	0.50	0.75	0.38	0.50	0.75

Table 3: Reliability indicators and weights for modified IEEE 14 bus network [1/2]

transformer is interrupted, the power flow will continue). Therefore, a substation is only constructed with 1 transformer, with one exception of a substation with four different voltages. For that substation, the assumption of having two transformers gives a more effective distribution. The data for this table is provide by ABB [14]. Transformer weight is based on the conductor material (copper) weight in the wiring and the profile during the manufacturing phase. Furthermore, the transformers under evaluation are assumed to be almost equal to the ones from the manufacturer, e.g. for a 110/10 kV substation, 220/15.6 kV transformer is used [110 –220]. For a network substation, the different combinations of switch-gear are assumed based on their operating voltage level. These combinations are needed for environmental impact assessment, which will be talked about later on. Switch-gear values are based on the environmental impact per transformer.

2.2 Node reliability indicators

Given reliability indicators combine the factors related to outage duration or the response time, frequency of outage, number of customers involved in interruption or their lost power and energy. System average interruption frequency index (SAIFI), system average interruption duration index (SAIDI), energy not supplied (ENS), average service availability index (Availability), average interruption time (AIT), average interruption frequency (AIF), average interruption duration (AID) are among the measures to evaluate the interruptions and its impacts.

2.2.1 *Terrain and weather effects*

The failure rate of any component in a power system network is assumed to be influenced by different internal (quality and type of transmission line) and external factors (terrain, topology and weather conditions). Nine different terrain types are considered for this study that outlines the network environments. For example, the hill-like location has a lower failure rate compared to a commercial location, because in the commercial surrounding, there are more internal factors related to the consumption, and seaside has a lower failure rate than the forest, because forest is prone to fail e.g, trees falling. For the outage of a generation unit a maintenance cost incurs for the system operator. Factors like replacement or repair or maintenance cost for a corridor or line segment of the power network is considered. Additionally the environmental impact of an outage in terms of fuel consumption during the servicing and resulting environmental impact is also taken into consideration [15].

3 Evaluation of Power Network Condition under Outages

The fig. 2 presents the IEEE-14 bus network with zones. Each zone presents a geographical location. Tables 3 and 4 presents the indicators and the corresponding values (normalized) for measuring the system reliability based on nodes under the proposed framework.

The formulation for probability of failure Ω_n for bus n can be presented as in (1). In (2) frequency of failure Ω_n^f is presented [6].

$$\Omega_n = \sum_i [P(O_j)(P_{i,g}P_{l,i} - P_{g,i}P_{l,i})] \quad (1)$$

$$\Omega_n^f = \sum_i [O_j^f(P_{i,g}P_{l,i} - P_{g,i}P_{l,i})] \quad (2)$$

Nodes	8	9	10	11	12	13	14
Voltage level (kV)	110/35/6	110/10/6	110/35/10	110/10	110/10	110/35/6	35/6
Generation [0/1]	0.00	0.00	1.00	1.00	0.00	0.00	0.00
Generation, kWh	0.00	0.00	2000.00	1000.00	0.00	0.00	0.00
Transformer [0,1,2]	1.00	1.00	2.00	1.00	1.00	1.00	1.00
Transformer type[1,2,3]	2.00	2.00	1+3	1.00	1.00	2.00	3.00
Life expectancy	0.20	0.20	0.20	0.20	0.20	0.20	0.20
Conductor weight (kg/Trfo)	0.57	0.57	1.00	0.61	0.61	0.57	0.00
Conductor weight (kg/kVA)	0.00	0.00	1.00	1.00	1.00	0.00	0.00
Trfo_oil weight (kg/Trfo)	0.22	0.22	1.00	0.56	0.56	0.22	0.00
Trfo_oil weight (kg/kVA)	0.00	0.00	1.00	1.00	1.00	0.00	0.00
Energy demand (kWh)	325.00	350.00	275.00	300.00	300.00	325.00	300.00
Energy demand	0.75	1.00	0.25	0.50	0.50	0.75	0.50
Switch-gear type	2.00	2.00	4.00	1.00	1.00	2.00	3.00
Life expectancy	0.00	0.00	0.00	0.00	0.00	0.00	0.00
GWP	0.02	0.02	1.00	0.14	0.14	0.02	0.00
AP	0.03	0.03	1.00	0.22	0.22	0.03	0.00
NP	0.05	0.05	1.00	0.44	0.44	0.05	0.00
SF6 (% of all emissions)	0.11	0.11	1.00	0.00	0.00	0.11	0.02
Total value:	0.05	0.05	1.00	0.15	0.15	0.05	0.01
Terrain	7.00	5.00	9.00	5.00	1.00	4.00	5.00
Terrain coefficients	0.75	0.50	1.00	0.50	1.00	0.38	0.50

Table 4: Reliability indicators and values for modified IEEE 14 bus network [2/2]

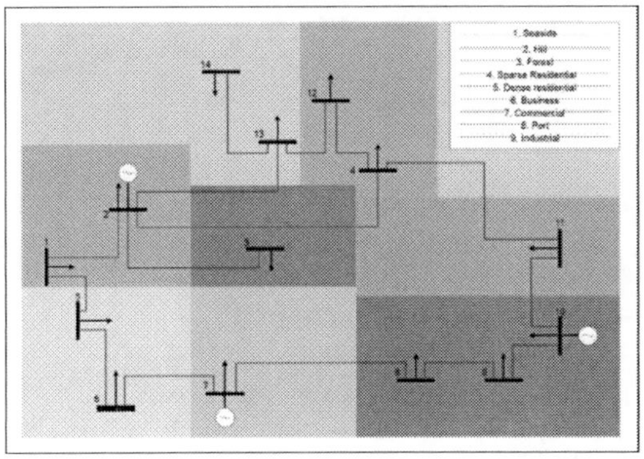

Figure 2: Classification of modified IEEE-14 bus network into zones based on weights

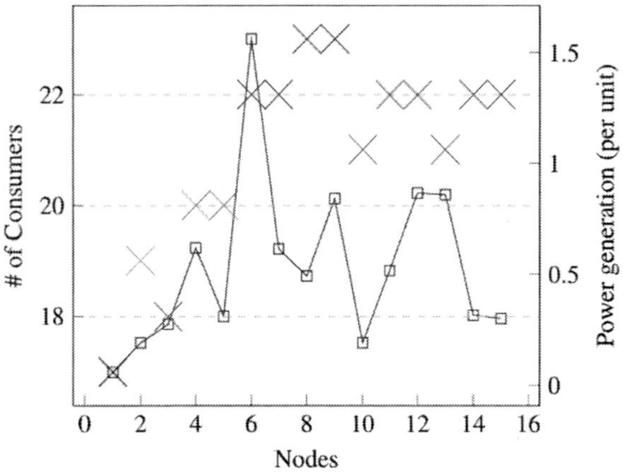

Figure 3: Node wise number of consumers with respective consumption in per unit

Where O_j is the condition of outage in the power transmission network. $P_{i,g}$ is the probability of occurrence of capacity outage beyond reserves. And probability of uninterrupted power supply. The availability (γ) is calculated as $\gamma = 1 - \frac{60*ENS}{\sum_i P_i}$ where P_i is average power supplied by the total system and ENS (Energy not supplied because of interruption) and P_i stands for power interruption for incident i. The total cost is product of component capital cost times availability. And the repair cost is the cost of repair times the total cost. Similarly the maintenance cost is value of fault times the repair costs.

Considering the four generation units 15 generation outage states are assessed. Figure 3 demonstrates the network topology and number of consumers with respective per unit energy consumption. There are 314 numbers of consumers in the 14 bus network with total consumption of 8.008. Table 5 states the power system indicators for evaluating the faults and losses with respect to line for interrupted load. The sum total SAIFI, SAIDI, ENS for the network are 0.00426, 8.925 and 0.196 respectively. The relative cost in € for investment, repair and maintenance are presented in table 6. The net overall investment is 29.29215 followed by 0.124837 for repair and 0.0005320 for maintenance cost.

Line	Interrupted load	SAIFI	SAIDI	ENS	Availability	CAIDI	CAIFI
1	0.0002514	0.000231	0.48	0.001442673	0.999999849	0.226468312	0.000230734
2	0.0008140	0.000258	0.54	0.004670347	0.999999831	0.253111642	0.000257879
3	0.0011720	0.000244	0.51	0.006724321	0.99999984	0.239789977	0.000244307
4	0.0026381	0.000271	0.57	0.015135836	0.999999822	0.266433308	0.000271452
5	0.0013212	0.000271	0.57	0.007580144	0.999999822	0.266433308	0.000271452
6	0.0066484	0.000299	0.63	0.038145242	0.999999804	0.293076639	0.000298597
7	0.0026210	0.000299	0.63	0.015038028	0.999999804	0.293076639	0.000298597
8	0.0020925	0.000312	0.65	0.012005970	0.999999796	0.306398304	0.00031217
9	0.0035842	0.000312	0.65	0.020564198	0.999999796	0.306398304	0.00031217
10	0.0008140	0.000285	0.60	0.004670347	0.999999813	0.279754973	0.000285025
11	0.0021991	0.000299	0.63	0.012617272	0.999999804	0.293076639	0.000298597
12	0.0036907	0.000299	0.63	0.021175500	0.999999804	0.293076639	0.000298597
13	0.0036609	0.000285	0.60	0.021004335	0.999999813	0.279754973	0.000285025
14	0.0013425	0.000299	0.63	0.007702405	0.999999804	0.293076639	0.000298597
15	0.0012785	0.000299	0.63	0.007335623	0.999999804	0.293076639	0.000298597

Table 5: indicators regarding system interruptions and losses10^{-3}

Line	Loss			Cost		
	LOLE	*LOEE*	*EIR*	*Investment*	*Repair*	*Maintenance*
1	0.001051799	1.85256E-06	0.999998147	3.49999	0.014916260	0.000063570
2	0.003404977	1.94149E-05	0.999980585	3.49993	0.014915998	0.000063569
3	0.004902454	4.0247E-05	0.999959753	0.59998	0.002556975	0.000010897
4	0.011034979	0.000203915	0.999796085	0.59988	0.002556556	0.000010896
5	0.005526403	5.11437E-05	0.999948856	0.59997	0.002556947	0.000010897
6	0.027810287	0.001295143	0.998704857	3.49547	0.014896969	0.000063488
7	0.010963671	0.000201288	0.999798712	0.59988	0.002556563	0.000010896
8	0.00875311	0.000128301	0.999871699	3.49955	0.014914374	0.000063562
9	0.014992597	0.00037641	0.99962359	3.49868	0.014910673	0.000063546
10	0.003404977	1.94149E-05	0.999980585	0.59999	0.002557028	0.000010898
11	0.009198787	0.000141699	0.999858301	3.49950	0.014914174	0.000063561
12	0.015438275	0.000399121	0.999600879	0.59976	0.002556057	0.000010893
13	0.015313485	0.000392695	0.999607305	0.59976	0.002556074	0.000010893
14	0.005615539	5.28068E-05	0.999947193	0.59997	0.002556943	0.000010897
15	0.005348132	4.78973E-05	0.999952103	3.49983	0.014915573	0.000063567

Table 6: Relative costs for lines (€)

4 Conclusion

This study is about reliability of power distribution network. The presented reliability framework includes interdisciplinary aspects of life cycle analysis, consumer satisfaction index and power interruptions. These measures are summarized in terms of investment, repair and maintenance costs in the presented framework. The case study conducted on modified IEEE-14 bus network for generation outage scenarios demonstrate the relation between investments, repairs and maintenance costs to maintain the reliability. Apart from that, the system operator has an incentive in form of reducing direct investment to with marginal repair or maintenance cost. It is evident from the case study that the outages has an adverse impact on the system reliability and therefore results in additional cost. However, the additional cost can be shared with timely maintenance and repairs to avoid an up-front investment. Therefore, in an investment decision for network expansion, reliability analysis of the power network becomes significantly important to make an optimal decision. In future works, investment models with network restructuring will be studied. Application of classification techniques to evaluate weights for each edge in another research avenue, given that there are both empirical and quantitative information is included in the reliability criterion.

References

[1] R. Billinton, R. Karki, IEEE Transactions on Power systems **14**, 1172 (1999)

[2] A.M.L. da Silva, L.A.F. Manso, W.S. Sales, L.C. Resende, M.J. Aguiar, M.A. Matos, J.A.P. Lopes, V. Miranda, European Transactions on Electrical Power **17**, 387 (2007)

[3] R. Karki, R. Billinton, A.K. Verma, *Reliability Modeling and Analysis of Smart Power Systems* (Springer, 2014)

[4] H. Farzin, M. Fotuhi-Firuzabad, M. Moeini-Aghtaie, IEEE Transactions on Power Systems **33**, 2359 (2018)

[5] R.N. Allan et al., *Reliability evaluation of power systems* (Springer Science & Business Media, 2013)

[6] R. Billinton, W. Li, in *Reliability Assessment of Electric Power systems Using Monte Carlo Methods* (Springer, 1994), pp. 9–31

[7] A. Volkanovski, M. Čepin, B. Mavko, Reliability Engineering & System Safety **94**, 1116 (2009)

[8] J. Pylvänäinen, P. Verho, J. Järvinen, S. Kunttu, J. Sarsama, *Advanced failure rate and distribution network reliability modelling as part of network planning software*, in *Electricity Distribution, 2005. CIRED 2005. 18th International Conference and Exhibition on* (IET, 2005), pp. 1–5

[9] P. Jarventausta, J. Partanen, S. Karkkainen (2008)

https://doi.org/10.1051/e3sconf/20198002005

[10] J. Pylvanainen, J. Jarvinen, P. Verho, *Advanced reliability analysis for distribution network*, in *Electric Utility Deregulation, Restructuring and Power Technologies, 2004.(DRPT 2004). Proceedings of the 2004 IEEE International Conference on* (IEEE, 2004), Vol. 2, pp. 457–462

[11] W. He, X. Shen, J. Guo, Q. Wang, Y. Lu, H. Li, *An improved design of damped AC test system for partial discharge measurement in distribution power cables*, in *IOP Conference Series: Materials Science and Engineering* (IOP Publishing, 2018), Vol. 366, p. 012029

[12] A. Cataliotti, A. Daidone, G. Tinè, IEEE Transactions on Power Delivery **23**, 1896 (2008)

[13] E. Cables, https://www.elandcables.com/electrical-cable-and-accessories/cables-by-type/medium-voltage-cable (2018)

[14] A. Cables, XLPE Land Cable Systems (2018)

[15] N. Ekstedt, P. Hilber, *Categorization and review of failure rate factors used in power systems*, in *Probabilistic Methods Applied to Power Systems (PMAPS), 2014 International Conference on* (IEEE, 2014), pp. 1–6

Analysis on building industry CO_2 emission in Shanghai

Hao Guo and *Qingcheng* Wang

College of Urban Construction and Safety Engineering, Shanghai Institute of Technology, 201418 Shanghai, P.R.China

Abstract. With the economic development in Shanghai, a lot of energy has been consumed in the building industry, and is accompanied by a great amount of carbon emissions. Based on the data of energy consumption in the two stages of the building materials production and construction process in Shanghai, a calculation model of CO_2 emissions was created, total CO_2 emissions have been evaluated during recent ten years, and CO_2 emissions influencing factors was identified in each stage. The research results show that CO_2 emissions in the stage of building materials production, account for more than 90% of total CO_2 emissions, mainly traced from cement, iron and steel industry, and have greater potential for energy conservation and emissions reduction.

1 Introduction

With the rapid development of urbanization in China, the total of building energy consumption and the intensity of energy consumption continues being increasing, greenhouse gas emissions are also rising. According to statistics, about 40% of the energy consumed in the world each year are related to the building industry, and CO_2 emissions from the construction sector account for about 36% of global carbon emissions[1]. It is anxious to reduce carbon emissions in the building industry. In 2016, the State Council issued the "Thirteenth Five-Year Plan" for the control of greenhouse gas emissions, and further proposed the total energy consumption and carbon emissions control targets: By 2020, the CO_2 emissions per unit of GDP will decline by 18% from 2015, and the total carbon emissions will be controlled effectively. The CO_2 emissions calculation model is established, which is an essential foundation for reducing carbon emissions and analyzing the CO_2 emissions in the building industry so as to implement targeted carbon emissions reduction measures.

In recent years, some results have been obtained since scholars did relevant researches on building carbon emission. In the past studies, some literature research on CO_2 emissions from the life cycle of the building of all stages, including building materials mining and production, construction, building use and maintenance and recycling stage [2]. Some literature compares CO_2 emissions from different building types and analyses it. However, from a macro perspective, only a few studies have been conducted on total CO_2 emissions during the building materials production and construction process stage.

In this paper, the calculation boundary of CO_2 emissions in the building industry is determined firstly, following available data, concise calculation model and the scientific calculation method. Total CO_2 emissions can be divided into two stages: the building materials production and construction process. Calculation model of CO_2 emissions has been created and emission factors have been identified. CO_2 emissions of the building industry have been evaluated in Shanghai yearly.

Key factors affecting carbon emissions are to be identified and basic data are to be provided for energy conservation and emission reduction in the building industry in Shanghai.

2 Study method

2.1 Carbon Emission Factor

Carbon Emission Factor, which is an important parameter used to signify each kind emission characteristics of energy of greenhouse gas, indicates the production of CO_2 accompanied by energy consumption per unit of mass [3]. The carbon emission factor is the unit quality of sum of the amount of greenhouse gas emitted from each phase to CO_2. The International Energy Agency (IEA) unifies the emissions of various types of greenhouse gas into carbon dioxide equivalent (CO_2-eq) for quantitative analysis of carbon emissions[4].

2.2 Carbon Emission Calculation Boundary and Stage

Due to the depth of research and also inaccessibility of data, it cannot be fully calculated the CO_2 emissions from two stages of building materials production and construction process. Thus, in building materials production stage, four main materials are regard as calculation objects, and the total energy consumption is used as to calculate in construction process stage.

3 Carbon emission calculation model

CO_2 emissions of the building industry mainly come from two major stages, the building materials production and the construction process. The carbon emissions model proposed being based on "Volume = activity data × emissions". After obtaining relevant data from sub-activities, energy consumption of sub-fuel species and corresponding emission factors, the final summation will be done [5], the calculation model is as following.

$$E_c = E_{cm} + E_{ce} \quad (1)$$

Comments: E_c indicates the total of CO_2 emissions. E_{cm} indicates the CO_2 emissions in the stage of building materials production. E_{ce} indicates the CO_2 emissions in the construction process.

3.1 Calculation Model of CO_2 Emission in the Stage of Building Materials Production

There are many types of building materials, due to the difficulty of time and statistics, it is impossible to count all kinds of building materials completely, and the carbon emission factors of materials are affected by many aspects, such as the scale of building materials manufacturers and production technology. The calculation of CO_2 emissions in building materials must take into consideration the recyclable factor of building materials. Although glass and wood can be recycled in part, the recycled glass is no longer part used in building generally. Therefore, glass and wood recyclable factors are not considered in calculation. Recyclable factor of renewable materials has shown in Table 1.

Table 1. Recyclable materials recovery factor[6].

Materials	Factor
Steel	0.35
Rebar	0.40
Aluminum	0.95
Plate glass	0.80
Wood	0.10

The CO_2 production stage of building materials is mainly from the energy consumption (coal, diesel, etc.), the decomposition of raw materials (decomposition of limestone, etc.), and the energy workers (energy, etc.). The calculation process can be performed according to the formula (2).

$$E_{cm} = \sum_{i}^{n} p_i \times q_i \times (1 - \alpha_i) \quad (2)$$

Comments: α_i indicates the recovery factor of building materials. p_i indicates the emission factor of building materials. q_i indicates the consumption of building materials.

Table 2. The consumption of major building materials in Shanghai over the years.

Year	Cement (10^8 t)	Steel (10^8 t)	Aluminum (10^4 t)	Glass (10^4 t)
2006	19.10	8.87	16.17	58.70
2007	18.88	9.74	12.39	53.70
2008	19.26	11.26	15.67	58.90
2009	20.62	12.63	17.59	64.30
2010	21.63	13.65	19.29	65.30
2011	23.25	14.63	20.96	83.20
2012	21.12	15.23	52.52	98.10
2013	21.58	15.80	37.76	53.70
2014	22.86	16.78	40.12	95.10
2015	17.39	15.08	36.49	93.50

Table 3. Carbon emission factor in major building materials production stage.

Materials	Unit	Emission factor
Rebar	tCO_2/t	2.03
Cement	tCO_2/t	0.70
Aluminum	tCO_2/t	1.22
Glass	tCO_2/t	1.40

Steel is one of the important materials for building production. Its carbon emissions are related to the production process closely. In the steel-making process, carbon emissions are primarily caused by energy consumption and combustion. Although steel is a high-carbon emissions building material, the recovery rate of steel is high relatively. Therefore, the degree of steel impact on the environment needs to consider the steel recycling rate generally. Rebar, which is recovery in

 https://doi.org/10.1051/e3sconf/20198003001

reinforced concrete, is relative difficult and the recovery factor is 0.4. The recovery rate of Merchant steel and Die steel is superior, the recovery factor can be as high as 0.9, average calculation is taken as an average of 0.4.

There are three main cement manufacturing processes: wet rotary kiln, vertical kiln and new dry process. Due to the increasing proportion of new dry process, the results of the new dry process technology are used as reference. According to relevant studies, 50% of carbon emissions come from carbonate decomposition during cement production, 40% from fuel calcinations, and indirect emissions account for 10% of total cement carbon emissions [7]. Therefore, only the production of cement is regard as the calculation directly.

3.2 CO_2 emission calculation model during construction stage

From the perspective of energy statistics system in China, the energy consumption of new construction, existing building alteration. And building demolition is all included in the energy consumption of building industry. In addition, transportation energy consumption of the building materials and construction waste are related to construction activities. It also included in the energy consumption of the building industry. Therefore, the energy consumption of the building industry is utilized to signify energy consumption in construction process[8], and energy consumption is calculated as the formula (3).

Depending on the Energy Balance Sheet from "China Energy Statistical Yearbook" of the past years, "Shanghai Statistical Yearbook", "Shanghai Energy and Environment Statistical Yearbook in 2016", and "Shanghai Industrial Energy Transportation Statistical Yearbook in 2009", the energy consumption of major building materials during 2006 - 2015 is 1.72, 1.79, 1.85, 2.11, 2.17, 2.11, 2.13, 2.14, 2.06, and 2.18 million tce.

The calculated energy consumption is summarized and converted according to the carbon emission factors of fuel type used. Finally, the CO_2 emissions during the construction process are calculated, as the formula (3). According to relevant literature, the CO_2 emissions coefficient (unit: tc /tce) of the complete combustion of 1 ton standard coal: The National Energy Research Institute of the National Development and Reform Commission recommends value is 0.67. The value is 0.68 in Japan Institute of Energy Economics, and the reference value of the US Department of Energy Information Agency is 0.69. This paper calculates value is 0.68.

$$Q_c = Q_{ct} + Q_{ce} = \sum_i n_i k_i + \sum_j m_j k_j \quad (3)$$

$$E_c = Q_c \times \beta \quad (4)$$

Comments: Q_c indicates energy consumption during construction. Q_{ct} indicates energy consumption during transportation. Q_{ce} indicates energy consumption during construction process. n_i indicates consumption of fuel for transportation. k_i indicates standard coal conversion factor of fuel for transportation. m_j indicates the energy consumption of construction equipment. k_j indicates the standard coal conversion factor of construction equipment. β indicates the standard coal carbon emission conversion factor.

As shown in Figure 1, the sum of emissions in building materials production and construction process stage indicate that the change of CO_2 emissions in the building industry of shanghai. The results indicate that CO_2 emissions increased from 33.56 to 46.02 million tons from 2006 to 2015.

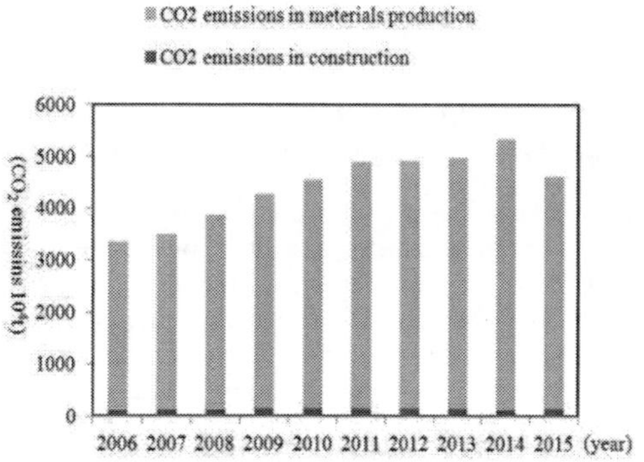

Fig. 1. Total CO_2 emissions over the years in Shanghai.

4 Results analysis

The result show that Shanghai has a large amount of energy for construction and CO_2 emissions. In 2015, the energy for building construction in Shanghai reached 14.84 million tce, accounting for 12.85% of total terminal energy consumption in Shanghai and 46.02 million tons CO_2 emissions for construction process. The rapid growth of energy consumption in building , as the primary industry, has led to the rapid growth of total energy consumption in recent years in Shanghai. Owing to the fact that continued growth of the construction scale in recent years, the production energy consumption of the unit products of building materials has been declining year by year along with the improvement of production technology. In 2015, Shanghai's housing construction area was 3660 million m^2, which was nearly 1 time higher than in 2006, as showed in the Figure 2. Comprehensive energy consumption per unit product of steel and cement decreased by 8.7% and 11% respectively compared to 2006. A continuous increase in the construction of energy consumption, and CO_2 emissions continue to rise on account of the growth rates of the construction scale far exceeds the speed of the decline in construction energy consumption.

CO_2 emissions per unit construction area ranged from 0.21 to 0.12 m^2/tons during 2006 - 2015. Because continuous optimization of construction technology and processes, upgrading of technical equipment and continuous innovation of technologies, as showed in the Figure 3.

 https://doi.org/10.1051/e3sconf/20198003001

It is significant that controlling construction energy consumption for achieving energy conservation and emission reduction. As China's urbanization construction enters a new phase, further changes have taken place in the urbanization of China. The construction of large-scale residential, buildings, and commercial buildings has come to an end. The problem of absolute shortage has also been eased. Therefore, China should gradually improve its infrastructure in the future, slow down its construction, and slightly reduce its basic construction scale.

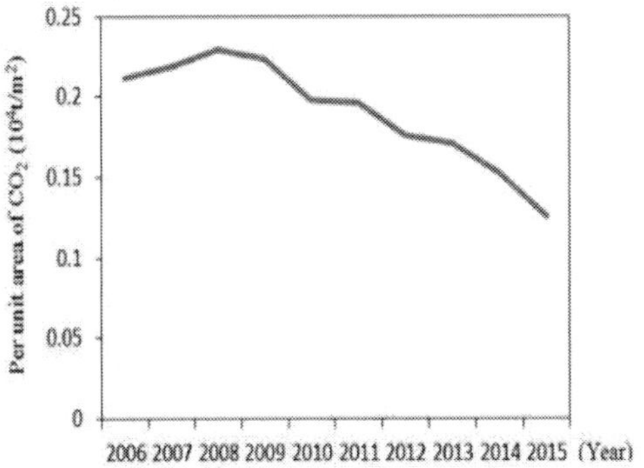

Fig. 2. Building construction area in Shanghai.

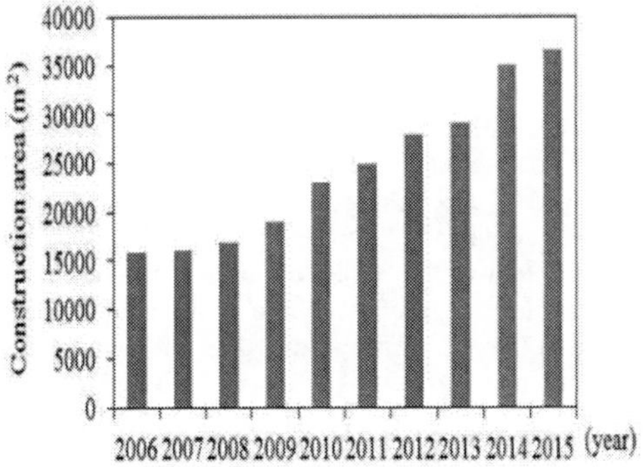

Fig. 3. Per unit area of construction carbon emissions.

5 Conclusion

The two stages of CO_2 emissions from the building industry have been calculated in this paper. The research results show that CO_2 emissions in the production stage of building materials are dominating, while the proportion of construction process stage is insignificant. The energy consumption of building materials in the cement industry, iron and steel industry account for a large proportion of CO_2 emissions, account for more than 90% of total CO_2 emissions, and there is a huge potential for energy conservation and emission reduction. Built on statistical data, the CO_2 emissions from the macro-building industry are calculated year by year. CO_2 emissions controlled in the building industry will be greatly significant to achieve the goal of energy conservation and CO_2 emission reduction in China. The scale of construction should be controlled, and the building construction speed should be reduced also, and the energy saving technology should be employed to reduce the CO_2 emissions.

References

1. Y Gao, Z Li, H Zhang, B Yu, J Wang. A Carbon Emission Analysis Model for Prefabricated Construction Based on LCA. Journal of Engineering Management. **2**,30-34(2008).
2. S Zhang, W Xu, D Sun.Study on the determination and application of calculation methods for building carbon emissions. Journal of Building Science. **29**,35-41(2013).
3. Change I C. IPCC Fifth Assessment Synthesis Report. UK New York, Cambridge University(2014).
4. R Mao. Research on Environmental Impact Analysis of Urban Transportation Infrastructure Based on LCA. PhD dissertation. Shenzhen University(2017).
5. B Li. Emission estimation method. PhD dissertation. Wuhan: Huazhong University of Science and Technology(2012).
6. X Wei, Q Lai, X Li. J Qiao. The Efficient Energy Consumption of Mechanical and Electrical Energy Consumption in Residential Buildings. Journal of Beihua University.**4**,484-487(2013).
7. M Zhao, J Cheng, J Hu. The accounting of CO2 emissions based on the energy balances, Journal of Ecological Economy.**11**,30-32+157(2012).
8. L Lin, Y Jiang, D Yan, Y Peng. Analysis of General Construction Energy Consumption and CO2 Emission in China Construction Industry. Journal of Expert Analysis.**3**,5-10(2015).
9. Y Wang. Whole life cycle carbon emissions research of industrialized precast construction. PhD dissertation. Southeast University(2015).

 https://doi.org/10.1051/e3sconf/20198003002

Characterization of wastewater from landfills – amount and type of humic substances extracted from leachate

Maria Elisabete Silva[1,2], *Marlene* Santos[3], and *Isabel* Brás[1]

[1]Departamento de Ambiente, Escola Superior de Tecnologia e Gestão de Viseu, Campus Politécnico, CI&DETS, Instituto Politécnico de Viseu, 3504-510 Viseu, Portugal
[2]LEPABE, Laboratório de Engenharia de Processos, Ambiente, Biotecnologia e Energia, Faculdade de Engenharia, Universidade do Porto, Portugal
[3]Departamento de Ambiente, Escola Superior de Tecnologia e Gestão de Viseu, Campus Politécnico, 3504-510 Viseu, Portugal

Abstract. This study aimed to characterize the humic substances (HS) extracted from landfills wastewater – leachates, with different ages of exploration. To reach the objective it was applied spectroscopy techniques, UV-Vis and FTIR spectra, as well as the ratio between the absorbance analysed. First, the HS were extracted, then fractionated in fulvic acids (FA) and humic acids (HA) and it was evaluated the phytotoxicity. HS content in the leachates were higher than the typical values found in the natural aquatic humic sources. It has been identified that the leachate HS, HA and FA aromatic fractions increased with the increase of the landfilling age, suggesting that the degree of humification increased with the landfilling age. All the HS showed a high aromaticity and humification degree. The HS extracts irrespective of their source presents similar structural composition. The functional groups found are in agreement with the literature: phenols, alcohols, carboxylic groups, aliphatic structures, among others. It was found that HA are mainly organic matter with a higher aromatic degree than FA. The HS and HA showed absence of phytotoxicity, testing by germination index, suggesting that may be used to produce liquid organic fertilizers.

1 Introduction

The disposal of solid wastes in sanitary landfills represents nowadays the most common method of waste management. Leachates or wastewater produced in landfills are generated mainly due to rainwater percolation through wastes and their decomposition products [1], leading to a complex mixture of high strength organic and inorganic contaminants [2]. Moreover, the recalcitrant organic contaminants may include humic substances (HS), phthalic esters, pesticides, and many other emerging organic pollutants [3]. The refractory HS include humic acids (HA), fulvic acids (FA) and humin that are produced during the decomposition of the biodegradable fraction of organic compounds in leachates. These compounds have a major role as fertilizers and land correctives, which may contribute to the essential functions of global soil fertility. In particular, the role of humic acids should be emphasized. For example, it addition to soil has a positive effect on the assimilation of nutrients by plants and microorganisms. Indeed, nowadays there is a growing tendency to use humic substances as liquid organic amendment, produced from natural resources such as leonardite, peat, and lignite [4, 5], but their use could be a contribute to the extinction of these natural resources. Thus, it is important to find other sources of these substances, such as the leachate. In general, HS are composed of chemically complex organic compounds, which are largely hydrophilic, acidic and polydisperse [6]. In nature, HS, mainly HA and humin, are extremely resistant to biodegradation. Indeed, given their low decomposition, HS belong to the group of stable natural organic compounds [7]. Despite great efforts to elucidate the still unknown HS structure, it is known that their major functional groups include carboxylic, phenolic, carbonyl, hydroxyl, amine, amide and aliphatic moieties, among others [8]. Briefly, HS are composed of a very large variety of aromatic rings (phenolic and quinonic) bound by acid functional groups or by peripheral aliphatic chains (polysaccharides, peptides, etc) and grouped into the different molecular arrangements that make up the HS structure [9].

In this study, an array of analyses was performed to characterize HS extracted from two landfills with different landfilling ages in order to chemically stablish their structural composition and related them with the landfill exploration period. Additionally it was evaluated their phytotoxicity

2 Materials and Methods

Three leachates samples from two Municipal Solid Waste (MSW) landfills - Bg and Pl - with different landfilling ages were used to carry out this study. They are located in the north of Portugal. Samples were collected at different operating stages corresponding to 12 years for landfill Bg and 16 and 17 years for Pl (samples code Pl_1 and Pl_2, respectively).

2.1 Humic Substances Extraction

Leachate extraction for humic substances (HS) was performed by the acid–base treatment method [10]. Briefly, leachate samples were filtered through 0.45 μm cellulose membrane filters and acidified to pH 2.0. Acidified leachate samples were percolated through a XAD-8 (Supelite DAX-8, Supelco) resin column. The preparative cleaning of the resin is described by Thurman and Malcom [11]. After percolating the whole sample through the adsorption column, HS were eluted from the XAD-8 resin using 0.1 M NaOH, in reverse flow. The HS fractionation in FA and HA was achieved by a selective precipitation process with pH control, using concentrate sulphuric acid. The resultant HS was fractionated by 1.5 mL 96% H_2SO_4 addition (pH 2). After 24 h at room temperature and centrifugation for 20 min at 4000 rpm, the insoluble fraction, which contained humic acid substances (HA fraction), was separated from the soluble fulvic acid substances (FA fraction). HA was dissolved in distilled water [12]. The total organic carbon (TOC) content of HS, HA and FA was measured in terms of mg TOC/L using a TOC-L CSN analyzer (Shimadzu).

2.2 Humic Substances Chemical Characterization

The absorbance of the HS, HA and FA fractions were measured at 254 nm, 280 nm, 465 nm and 665 nm in a UV-VIS spectrometer (Jenway 6300 Spectrophotometer). The ratio of quantified absorbance E254/E465, E280/E465, E280/E665, E465/E665 were evaluated and interpreted.

The functional groups of samples were identified by Fourier transform infrared (FTIR) covering a wavenumber range of 400–4000 cm^{-1}. FTIR spectra were recorded on a Bruker Tensor 27 spectrophotometer using KBr pellets. The FTIR evolution was followed by establishing ratios between absorbance at: 1034 cm^{-1}, 1235 cm^{-1}, 1384 cm^{-1}, 1544 cm^{-1}, 1652 cm^{-1}, 2925 cm^{-1}.

2.3 Phytotoxicity Tests

The phytotoxicity evaluation was conducted performing germination tests, using *Lepidium sativum* (water cress) as tested seed. This bioindicator was chosen due to its rapid growth and high sensibility to pollutants. All the tests were done using seven replicates with fifteen seeds, in Petri dishes with sterile filter paper humidified with 4 mL of the testing solution. Blank tests were also done. A leonardite based commercial liquid fertiliser (Humistar®) was included as control. The germination index (GI, %) was evaluated by equation 1.

$$GI = \frac{\overline{GS}_s \cdot \overline{RL}_s}{\overline{GS}_c \cdot \overline{RL}_c} * 100 \quad (1)$$

Where $\overline{GS}$ represents de average number of germinated seeds and $\overline{RL}$ the average root length (mm) of seeds in samples – s, and the blank – c tests.

3 Results and Discussion

3.1 Humic Substances Quantification

After extraction of HS from the landfills leachates and their fractionation it was possible to quantify the HS, HA and FA (Table 1). The HS amount was similar for all the leachates evaluated and was higher than 780 mg/L of total organic carbon. All the leachates analysed registered higher FA concentration than HA. These results were higher than the reported by Fan et al that ranged between 100 and 202 mg/L for FA and 72 and 101 mg/L for HA [13]. Moreover, the amount found of HS was greater than the reported for the natural aquatic resource of HS [4]. Overall, the data indicate that the leachate may be a source of HS, because the high carbon content of the samples.

Table 1. Quantification of HS, FA and HA obtained from the landfill leachates

Landfill Leachate	**HS** [mg TOC/L]	**FA** [mg TOC/L]	**HA** [mg TOC/L]
Bg	839	677	28
Pl_1	780	288	30
Pl_2	1311	966	131

Once, the FA concentrations values are higher than the HA, is possible to state that the landfills are not in the maturation phase [14]. These results suggest that the HS from landfill leachates were in an early stage of humification, once the degree of humification increase as the landfilling age increase.

3.2 UV-Vis Spectrum Analysis

UV absorption of HS, HA and FA is used to evaluate the condensation level of humic aromatic core indicating the maturation degree [15], and to analyse their molecular properties [16-18]. There are three important regions in the spectrum where exact absorbance is measured. Zbytniewski and Buszewski [14] enlightened that UV-Vis analysis of alkali extracts is generally based on the assumption that the absorbance at 260–280 nm is due to lignin and quinone moieties. The absorbance at 254 nm gives the total dissolved organic matter in the solution and the values at 280 nm show the organic matter aromaticity degree [19]. The absorbance at 465 nm reflects the organic material at the beginning of humification and the absorbance at 665 nm is referred to be indicative of strongly humified material with a high degree of aromatic and condensed groups [16]. Therefore, the absorbance ratios have been calculated

E254/E465, E280/E465, E280/E665, E465/E665, and may be used to determine the degree of maturation (humification) of organic matter [20]. The E254/E465 ratio point to the proportion of the dissolved total organic matter and the organic matter at the humification beginning, indicating the biodegradable organic matter at this point. The E280/E465 reflects the proportion between the lignin and quinone and the content of organic matter at beginning of humification. E280/E665 ratio denotes the relation between aromatic organic material and strongly humified material [14]. The E465/E665 ratio is used to indicate the degree of condensation and polymerization of the aromatic constituents, and can fulfill the function of humification index [21]. It can also be used as an indicator of molecular weight [22]. In Figure 1, these ratios are represented. From these data is possible to record that the ratio for HS, HA and FA have a similar trend for the three samples. The values obtained in increasing order were the following: E465/E665, E280/E465, E254/E465 and E280/E665 ratio. Indicating that the 3 fractions present a high degree of humification (E465/E665 low) and additionally a high degree of aromaticity, namely the possible presence of phenolic and carboxylic groups (high E280/E665). For all cases, HA have a lower ratio than the other humic structures (HS and FA).

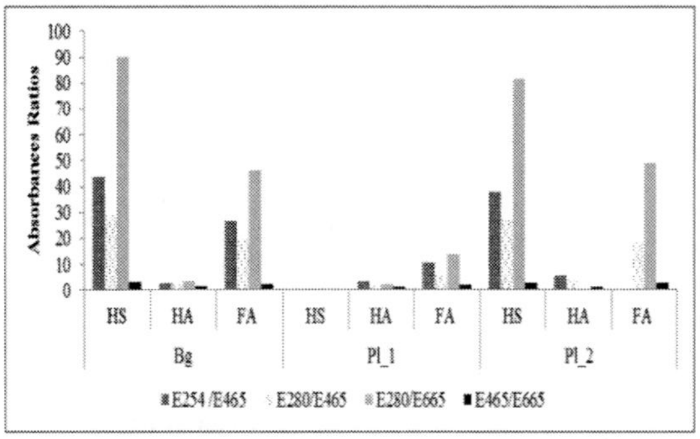

Fig. 1. Absorbance ratios of the humic substances extracted from the landfill leachates.

3.3 FTIR Spectrum Analysis

The changes in FTIR spectrum have been reported in the literature by calculating the ratio between the intensities of the major peaks absorption, this may be used as a tool for the study of humic structures evolution [25-26] and thereby the humification process. The FTIR absorbance evolution of humic structures was followed by establishing the ratios [17, 25-28]: 1652/2925 (aromatic C=C, COO^-/ aliphatic C-H); 1652/1544 (aromatic C=C, COO^-/ amide II N-H); 1652/1384 (aromatic C=C, COO^-/stretching N-O); 1652/1235 (aromatic C=C, COO^-/ aromatic ethers C-O-C, amide III); 1544/2925 (amide II N-H / aliphatic C-H); 1544/1384 (amide II N-H / stretching N-O); 1384/2925 (stretching N-O / aliphatic C-H); 1034/2925 (polysaccharides C-O/aliphatic C-H); 1034/1235 (C-O polysaccharide/ ethers aromatic, amide III C-O-C). These results are described in Table 2. The ratios changed between 0.96 and 1.50, values that were in accordance with those found by other authors that reported ratio values between 0.53 and 1.53 [25-26, 28]. The ratio data followed the increasing order: 1034/1235<1034/2925<1384/2925<1544/1384<1544/2925<1652/1384<1652/1543<1652/1235<1652/2925. This suggests that the 3 fractions are mainly constituted by aromatic C=C and COO^- than aliphatic C-H, because the high 1652/2925 (aromatic C/aliphatic C) ratio. Overall, similar values were found for the 3 fractions analysed (HS, HA and FA), however the HA fractions registered a slightly high value for this ratio, followed by HS and then by FA. These results are in accordance with the higher aromatic character of the HA than the other humic structures. The data from the ratios showed an increase mainly by the intensity of structures absorbing at 1652, 1544 and 1235 cm^{-1} with a relative decrease of absorbance at 2925 cm^{-1}. This suggests a richness of humic structures in aromatic compounds and a loss of readily available aliphatic compounds from these macromolecules, mainly polysaccharides, as result of their oxidation.

FTIR spectra of HA for leachate Pl_1 and Pl_2 samples are represented in Figure 2. According with the information reported by other authors [2, 18, 27] is possible to state that the HA extracted from these samples have in their constitution the following functional groups: O-H bonds, evidencing the presence of phenols, alcohols and carboxylic groups as can be seen by the absorption band at 3400-3200 cm^{-1} (stretching). The absorption spectrum for Pl_2 sample, from the older landfill has a stronger evidence of the carboxylic acids once the absorption band is shifted to lower values of wave numbers. The band at 3180-3090 cm^{-1} evidence the stretching of NH_2 bonds indicating the presence of primary amides. The band at 1640-1600 cm^{-1} represent C=O stretching bonds indicative of amides, carboxylates and aromatic rings; the band at 1140-1080 cm^{-1} represent the stretching vibrations from links of S-O relative to inorganic sulfates and the band at 750-700 cm^{-1} suggest N-H bonds indicative of secondary amine groups. It is also interesting to notice the band comprised between 2100-2000 cm^{-1}, which indicates the presence of cyanide ions, that it is also not common in this kind of substances [29].

In general, it is possible to conclude that the functional groups observed are in agreement with those registered in the literature [2, 27], therefore with the molecular characterization of substances.

Table 2. Absorbance ratio in mid-infrared measured by FTIR.

		1652/2925	1652/1544	1652/1384	1652/1235	1544/2925	1544/1384	1384/2925	1034/2925	1034/1235
Bg	**HS**	1.14	1.02	1.08	1.04	1.13	1.06	1.06	1.07	0.98
	HA	1.43	1.27	1.31	1.31	1.13	1.03	1.09	1.11	1.01
	FA	1.32	1.15	1.20	1.19	1.15	1.04	1.10	1.15	1.03
Pl_1	**HS**	1.33	1.17	1.20	1.23	1.14	1.03	1.11	1.09	1.00
	HA	1.50	1.41	1.45	1.47	1.06	1.03	1.03	1.02	0.99
	FA	1.28	1.14	1.19	1.19	1.12	1.04	1.08	1.09	1.01
Pl_2	**HS**	1.12	1.00	1.01	1.07	1.12	1.02	1.10	1.02	0.97
	HA	1.06	1.03	1.01	1.02	1.03	0.98	1.05	0.99	0.96
	FA	1.03	1.11	1.08	1.06	0.93	0.98	0.95	0.95	0.98

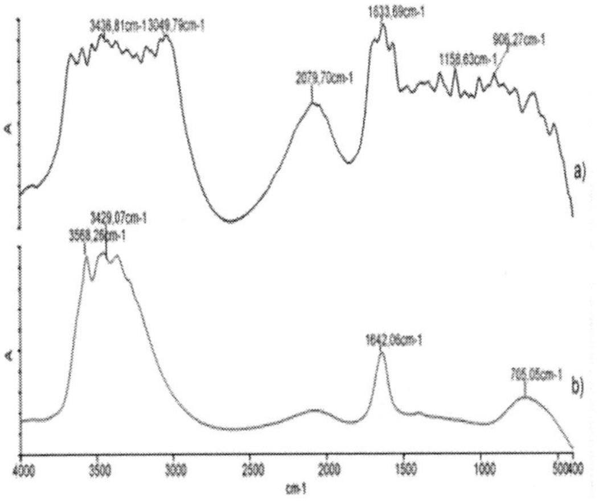

Fig. 2. FTIR spectra of AH for Pl_2 (a) and Pl_1 (b).

3.4 Germination Index

The HS, mainly the HA have several applications, such as fertilizers and land correctives, and may be applied in soils however its toxicity must be previously evaluated. The extracts from Pl_2 were diluted to achieve low electric conductivity (<4mS/cm) and were registered germination index of 115% for HS and 68% for HA. These GI values were lower than the value obtained for HS commercial liquid fertilizer Humistar® that reached 154%. It is possible to said that HS and HA showed absence of phytotoxicity because the GI are greater than 60% [30] and confirmed that the humic susbtances have a positive effect in the plant growth [31]. Overall, it seems that leachate may be used to produce liquid organic fertilizers an important way to valuing this type of waste.

4 Conclusions

In this work, it has been extracted HS and fractionated in humic and fulvic acids from two landfills with different landfilling ages and these fractions were characterized. The HS content in the leachates was higher than the typical values found in the natural aquatic humic sources. It was identified that the leachate HS, HA and FA aromatic components increased with increasing landfilling age. These results suggest that the degree of humification increased as the landfilling age increased. All the HS showed a high aromaticity and humification degree. The HS extracts regardless of their source presents similar structural composition. The functional groups found are in agreement with the literature: phenols, alcohols, carboxylic groups, aliphatic structures, among others. It was found that HA are constituted by organic matter with a higher aromatic degree than FA. The HS and HA showed absence of phytotoxicity, suggesting that may be used to produce liquid organic fertilizers.

The authors would like to express their gratitude to Instituto Politécnico de Viseu, the Center for Studies in Education, Technologies and Health (CI&DETS) and the Portuguese Foundation for Science and Technology (FCT).

References

1. S. Renou, J.G. Givaudan, S. Poulain, F. Dirassouyan, P. Moulin, Journal of Hazardous Materials **150**, 468-493 (2008)
2. L. Zhang, A. Li, Y. Lu, L. Yan, S. Zhong, C. Deng, Waste Management **29**, 1035-1040 (2009)
3. T.F.C.V. Silva, E. Silva, A.C. Cunha-Queda, A. Fonseca, I. Saraiva, M.A. Sousa, C. Gonçalves, M.C. Alpendurada, R.A.R. Boaventura, V.P.G. Vilar, Water Research **47**, 6167–6186 (2013)
4. IHSS, International Humic Substances Society, Available at: <http:// http://humic-substances.org/source-materials-for-ihss-samples/ > (2017) (accessed 10.05.18)
5. M.M. Tahir, M. Khurshid, M.Z. Khan, M.Z. Abbasi, M.H. Kazmi, Pedosphere **21**, 124 (2011)
6. P. Janos, Journal of Chromatography A **983**, 1–18 (2003)

7. P. Linnik, L. Malinovskaya, I. Zubenko, A. Zubko, Proc. Int. *Conf. on International Humic Substances Society - From Molecular Understanding to Innovative Applications of Humic Substances*, ed. Perminova, I.V. and N.A. Kulikova (Moscow) 427–430 (2008)
8. E.M. Peña-Mendez, J. Havel, J. Patocka, Journal of Applied Biomedicine **3,** 13–24 (2005)
9. A. Jouraiphy, S. Amir, P. Winterton, M. Gharous, J-C. Revel, M. Hafidi, Bioresource Technology **99,** 1066–1072 (2008)
10. American Public Health Association (APHA), Standard *methods for examination of water and* wastewater, 22th APHA, USA (2012)
11. E.M. Thurman, R.L. Malcom, Environmental Science &Technology **15**, 463–466 (1981)
12. M.E.F. Silva, L.T. Lemos, M.M.S. M. Bastos, O.C. Nunes, A.C. Cunha-Queda, Bioresource Technology **128**, 624–632 (2013)
13. H-J. Fan, H-Y. Shu, H-S Yang, W-C. Chen, Science of the Total Environment **361** 25 (2006)
14. R. Zbytniewski, B. Buszewski, Bioresource Technology **96,** 471 (2005)
15. J.W.C. Wong, K.F. Mak, N.W. Chan, A. Lam, A. Fang, L. X. Zhou, Q.T. Wu, X.D. Liao, Bioresource Techonology 76, 99-106 (2001)
16. E. Gieguzynska, A. Kocmit, D. Golebiewska, Humic *Substances in Ecosystems,* A. Zaujec, P. Bielek, S.S. Gonet, (Eds) (Nitra: Slovak Agricultural University) 35-41 (1998)
17. Z. Li, P. Kechen, P. Yongzhen, Bioresource Technology **250**, 413-421 (2018)
18. Z. Liu, W. Wu, P. Shi, J. Guo, J. Cheng, Waste Management **41**, 111-118 (2015)
19. J.L. Weishaar, G. Aiken, B.A. Bergamaschi, M. Fram, R. Fujii, K. Mopper, Environmental Science & Technology **37**, 4702 (2003)
20. R. Albrecht, J. Le Petit, G. Terrom, C. Périssol, Bioresource Technology **102**, 4495–4500 (2011)
21. H. El Hajjouji, N. Fakharedine, G. Ait Baddi, P. Winterton, J.R. Bailly, J.C. Revel, M. Hafidi, Bioresource Technology **98**, 3513–3520 (2007)
22. Y. Chen, N. Senesi, M. Schnitzer, Soil Science Society America Journal **41**, 352–358 (1977)
23. A. Lguirati, G. Ait Baddi, A. El Mousadik, V. Gilard, J.C. Revel, M. Hafidi, International Biodeterioration and Biodegradation **56**, 8-16 (2005)
24. K-H Kang, H.S. Shin, H. Park, Water Research **36**, 4023-4032 (2002)
25. P. Castaldi, G. Alberti, R. Merella, P. Melis, Waste Management **25,** 209-213 (2005)
26. S. Amir, A. Jouraiphy, A. Meddich, M.E. Gharous, P. Winterton, M. Hafidi, Journal of Hazardous Materials **177**, 524–529 (2010)
27. E. Smidt, K. Meissl, Waste Management **27**, 268–276 (2007)
28. R. Abouelwafa, G.A. Baddi, S. Souabi, P. Winterton, J. Cegarra, M. Hafidi, Bioresource Technology **99**, 8571-8577 (2008)
29. J. Coates, *Interpretation of Infrared Spectra, A* Practical *Approach,* Coates Consulting, Newton, USA, (2012)
30. A. Pera, G. Vallini, S. Frassinetti, F. Cecchi, Environmental Technology **12**, 1137–1145 (1991)
31. M. Morozesk et al., Chemosphere **184**, 309-317 (2017)

 https://doi.org/10.1051/e3sconf/20198003003

The effect of air conditioners on occupants' thermal adaptive behaviours and wellbeing: advances and challenges

Yuxin Wu[1, 2], *Mengwei* Yuan[1, 2], *Zixuan* Wang[1, 2] and *Xinyi* Li[1, 2]

[1]Joint International Research Laboratory of Green Buildings and Built Environments (Ministry of Education), Chongqing University, Chongqing 400045, China
[2] National Centre for International Research of Low-carbon and Green Buildings (Ministry of Science and Technology), Chongqing University, Chongqing 400045, China

Abstract. To study the impact of air conditioners (AC) on occupants' behaviours and comfort, two comparison surveys were conducted during hot summer weather before and after air conditioners were installed in student dormitories at Chongqing University in 2008 and 2016, respectively. The results showed that after the ACs were installed, occupants' environmental satisfaction vote was notably higher irrespective of whether the equipment was used. The proportion of ACs used (PAU) in 2016 and the proportion of fans used (PFU) in 2008 were logistically fit with the outside temperatures, and these data displayed very similar trends. However, less variety in adaptive behaviours was evident after the ACs were installed. When ACs were used, lower proportions of windows were opened (13%) and higher proportions of occupants felt stuffy (54%), experienced draughts (38%), and reported sick building syndrome (SBS). This study provides scientific insight into the advances and problems caused by the popularization of ACs.

1 Introduction

Before the air conditioner (AC) was invented, human beings worldwide suffered from changing weather conditions and developed adaptive skills involving behavioural, physiological, and psychological traits [1]. With the popularization of the AC, indoor thermal conditions and quality of life dramatically improved, especially on hot summer days. However, several problems such as sick building syndrome (SBS) [2], the energy crisis [3], and global warming have been drawing increasing attention from the public and researchers [4, 5], and thus, related concerns pertaining to the impact of ACs on human wellbeing, health, and work efficiency have become hot topics.

Staying in an air-conditioned space all day negatively affects people's health because it deteriorates occupants' physical functions that enable them to adapt to the rhythm of the natural climate cycle [6]. Furthermore, other studies have shown that the inherent ability of occupants to combat thermal stress can be degenerated [7] and adaptive skills in terms of behavioural responses can be weakened [8]. Field studies [9] have demonstrated that the human body has a higher acceptance of the thermal environment in free-running (FR) buildings than in AC buildings, which is related to the concept of thermal adaption [10, 11]. Thermal adaption to the environment is not only good for health [12], but also contributes to energy conservation because of the wider range of accepted air temperatures [13].

Hypotheses about how behaviours affect thermal comfort and energy usage have attracted the attention of many scholars. Previous work [14-16] has found that the occurrence probabilities of occupants' adaptive actions depend on indoor/outdoor conditions and can be predicted by using logistic regression analyses. Besides, the mean outdoor air temperature of the foregoing night(s) was found to have a major impact on occupant behaviour in summer [17]. Moreover, standardized occupant behaviour profiles in energy simulation tools can be weak due to human interactions [18], and personal characteristics can have a significant impact on household decisions involving whether to use air-conditioning systems [19]. Overall, the probability of human behaviours is governed by some rules but also can be affected by many other factors.

This study focused on the behavioural and psychological responses of occupants before and after ACs became available in summer. The effect of ACs on human thermal adaption was revealed, and it is hoped that this work can serve as a scientific reference for climate responsive solutions to cooling methods that can be employed in healthy buildings.

2 Research methods

Chongqing is renowned for its hot summers, and it averages 25 days with an outside temperature of more than 35 °C in a typical year [20]. Thus, it has earned the nickname 'Furnaces' in China. Despite this, no ACs were installed in the student dormitories at Chongqing

https://doi.org/10.1051/e3sconf/20198003003

University until 2012. Before 2012, the only mechanical cooling equipment installed was ceiling air fans. This provided an opportunity for us to conduct a comparative study of subjects' (occupants) responses before and after the ACs were installed. Because of the age of the subjects, building structure, outside environment, and the fact that the interior fitment had not changed much over the years, this site was better than other types of buildings for a long-term study. Illustrations of the dormitories are shown in Figure 1.

The field studies were conducted in the summers of 2008 and 2016. The psychological wellbeing and behavioural responses of occupants to the thermal environment were investigated through a questionnaire sent to students living in the dormitories. During the eight years of the study, the building structure did not change, which means that the natural indoor and outdoor environment can be treated as not having changed much.

The first survey was conducted between 27 June and 8 September 2008 by delivering approximately 650 copies of the paper questionnaire during daytime, and 573 valid copies were collected in 9 days. The second survey was conducted between 24 June and 15 July 2016 by sending online questionnaires to smart phones, and 428 valid copies were collected in 15 days. The total sample sizes met the requirements for a 90% confidence coefficient with a sample error of less than 5% using simple random sampling. To avoid unacceptable errors, days for which less than 20 students responded were not used to analyze the probability of behavioural responses. The outside thermal parameters came from the Shapingba meteorological station of the China Meteorological Administration (Site No. 57516) [21], and the data were collected on sunny, cloudy, and rainy days during the surveys.

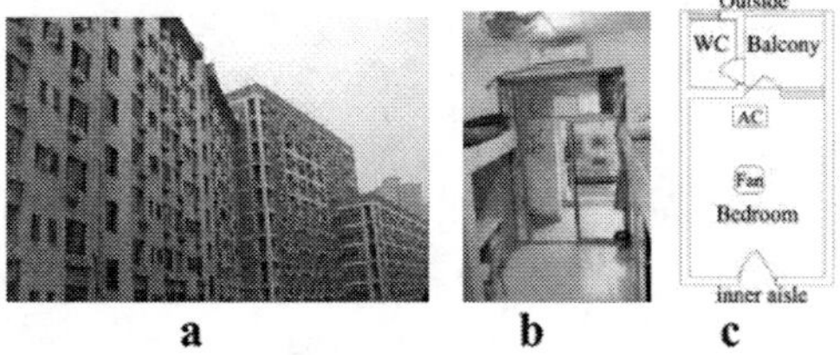

Fig. 1. Illustrations of the dormitories: a) Outside, b) inner part, and c) plane sketch.

The questionnaire was comprised of the following three parts: background information, subjective evaluation, and behavioural responses. Environmental satisfaction was evaluated on a five-point scale by asking 'How do you feel about the thermal environment in your dormitory: very satisfied (+2), satisfied (+1), indifferent (0), unsatisfied (-1), and very unsatisfied (-2)'. Thermal sensation was assessed by using the American Society of Heating, Refrigerating and Air-Conditioning Engineers (ASHRAE) seven-point thermal sensation scale [22]: cold (-3), cool (-2), slightly cool (-1), neutral (0), slightly warm (+1), warm (+2), and hot (+3). The SBS questions in this study were based on the 32 types of SBS recommended by the U.S. Environmental Protection Agency (EPA) [23], and questions were adjusted according to the actual situation.

During the same time, the use of ACs and fans and the opening of windows based on occupants' self-reports were recorded. Furthermore, the following question was asked to probe occupants' preferred behaviours: 'Which measurements would you take to improve indoor thermal conditions?'

3 Results and analysis

3.1 Occupants' wellbeing

3.1.1 Environmental satisfaction

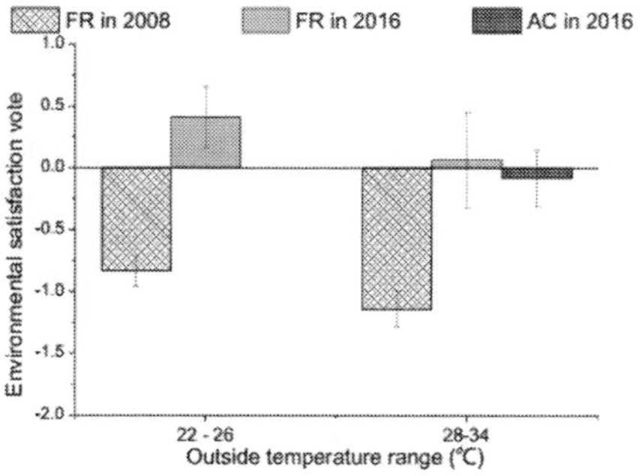

Fig. 2. Mean environmental satisfaction vote for different outside temperature ranges

Figure 2 shows the mean environmental satisfaction vote (ESV) for the following two outside temperature ranges: 22–26 °C (neutral) and 28–34 °C (warm). The ESV increased from about -1 (unsatisfied) to 0 (indifferent) after the ACs were made available in 2016. Moreover, the ESV was even better during neutral outside temperatures (22–26 °C) after the ACs became available in 2016 than that when ACs were not available in 2008. This means that the available ACs significantly improved occupants' wellbeing regardless of whether the equipment was used or not, which could be attributed to higher perceived control [24] and a better thermal experience [25].

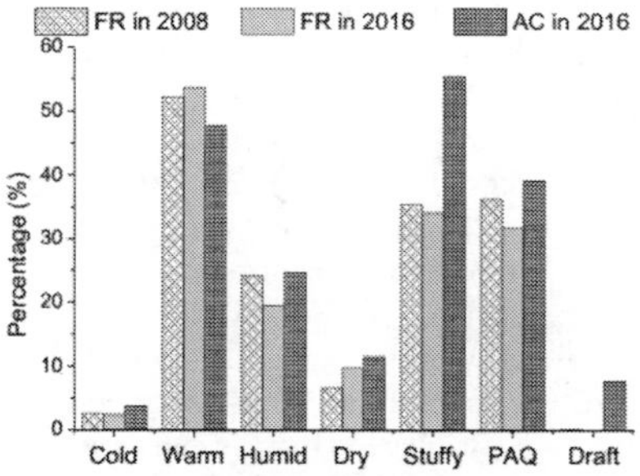

Fig. 3. Reasons for being unsatisfied

Figure 3 shows the main reasons for being unsatisfied with the environment, which included being too warm or stuffy, bad perceived air quality (PAQ), and high humidity (>20%) in all cases. Additionally, the data show that a higher proportion of occupants felt stuffy and experienced draughts when the ACs were used.

3.1.2 Thermal sensation vote

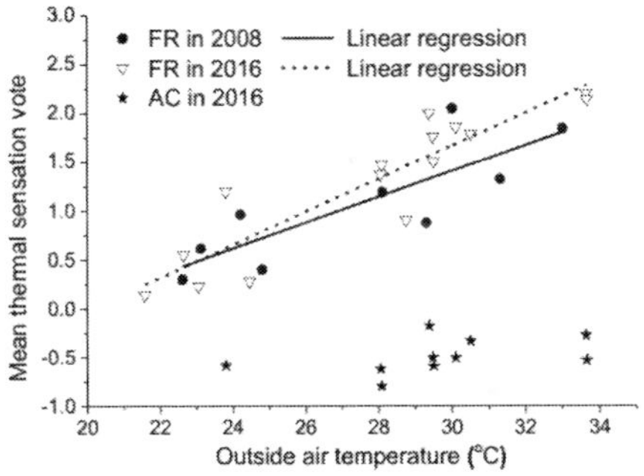

Fig. 4. The mean value of the TSV with the outside temperature

Figure 4 shows the thermal sensation vote (TSV) relative to the outside temperature. When ACs were used in 2016, occupants felt slightly cool and barely affected by the outside temperature. When ACs were not available in 2008 or not used in 2016, there were no significant differences in the TSV relative to the outside temperature.

3.1.3 Sick building syndrome

Table 1 provides the self-evaluation results for SBS during the survey. The prevalence of SBS was much lower among those who did not use the AC in the 2016 survey, which could imply that occupants who were not using the ACs were perhaps more accommodative to the warm thermal environment or that the environment created by AC use may increase the prevalence of SBS. Through an odds ratio (OR) analysis of the data in 2016, AC use was found to be a risk factor in the prevalence of SBS (OR = 1.184 (0.995, 1.409)), in which being absent-minded (OR = 1.53), eye discomfort (OR = 3.22), and discomfort in the upper respiratory tract (OR = 2.58) were significantly affected by AC use.

Table 1. The prevalence of SBS in 2016.

SBS	dizzy	nausea	fatigue	sleepy	absent-minded	eye	URY	skin	difficulty breathing
FR (%)	5.1	1.7	15.8	21.5	9.0	2.8	3.4	4.5	3.4
AC (%)	6.9	2.4	18.5	21.1	16.4	12.9	12.1	9.5	5.6
OR (AC)	1.22	1.31	1.12	0.99	1.53	3.22	2.58	1.67	1.39
P (χ^2)	0.45	0.54	0.47	0.93	0.03*	***	0.002**	0.06	0.29

Notes: *P <0.05, **P <0.01, ***P <0.001; URY – discomfort in the upper respiratory tract including dry lips, rhinobyon, or itchy throat; eye – discomfort including dry eyes, eyestrain, or lacrimation; skin – discomfort including dry skin, itchy skin, or skin disease.

3.2 Behavioural responses

3.2.1 Preferred behaviours

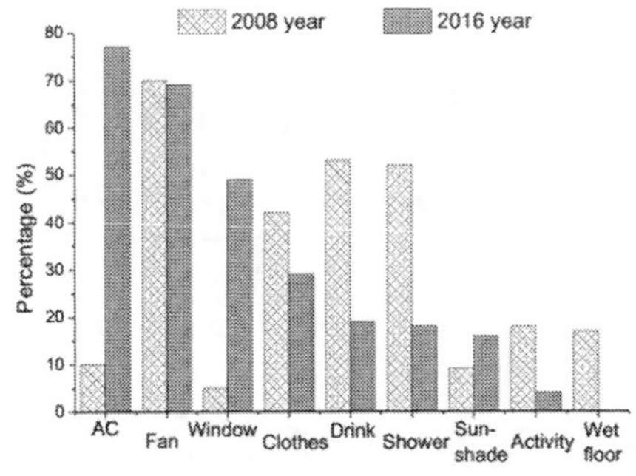

Fig. 5. Occupants' preferred behaviours

Occupants' preferred behaviours are shown in Figure 5. The most preferred behaviours in 2008 were using the fan, drinking something cold, showering, and adjusting clothing. As the window was always open, it was not a major adjustment measurement in 2008. In 2016, the most preferred behaviours were using the AC, using the fan, and adjusting the size of the window/door opening. The behaviours including drinking something cold, showering, and adjusting clothing were used less in 2016, and two behaviours, namely, sprinkling water on the floor and changing activities, were absent or low in 2016.

3.2.2 Use of mechanical equipment

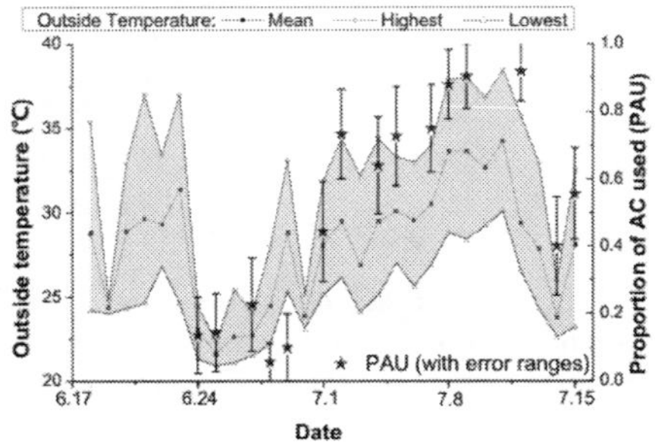

Fig. 6. Proportion of ACs used and outside temperature with time during the survey in 2016

Figure 6 illustrates the proportion of ACs used (PAU) relative to outside temperatures on different days during

https://doi.org/10.1051/e3sconf/20198003003

the survey in 2016. As shown, the PAU was highly related to the outside temperature. During the survey in the summer of 2016, the average temperature of the AC was set at 26 °C (41%), 25 °C (26%), and 24 °C (18%). Furthermore, the AC was mainly used at night and at noon.

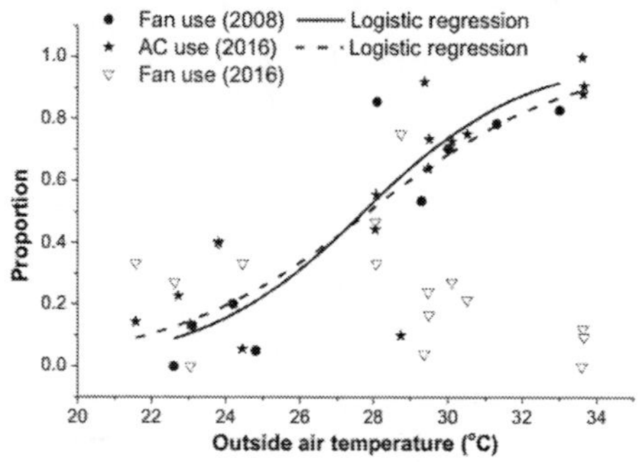

Fig. 7. Proportion of mechanical equipment used relative to outside temperature

Figure 7 shows the proportion of fans used (PFU) and PAU on different days alongside the outside temperatures in 2008 and 2016. Logistic regression analyses revealed that the relationships between the probability of a behaviour and outside temperature were as follows:
PFU in 2008:

$$PFU = e^{(-12.59+0.45t_{out})}/[1+e^{(-12.59+0.45t_{out})}];$$

$R^2 = 0.82$ (1)

PAU in 2016:

$$PAU = e^{(-10.19+0.37t_{out})}/[1+e^{(-10.19+0.37t_{out})}]\ ;$$

$R^2 = 0.72$ (2)

The PFU in 2016 did not change according to the outside temperature because the primary cooling method was an AC at that time in the summer. Therefore, only a proportion of occupants used the fan.

3.2.3 Non-mechanical adjustments

The typical clothing ensemble included short pants with a short sleeve shirt, with a mean clothing insulation value of about 0.3 clo. No significant difference was found for the different conditions.

Figure 8 shows the proportion of windows opened (PWO) in different conditions. The PWO was highest in 2008, when 91% of occupants opened the windows. When ACs were made available in 2016, the PWO was 82% during the free-running period, and it dropped to 13% after occupants started using the ACs. This was because the opening of windows when the AC was used would have caused more energy to be consumed and increased the electric bill.

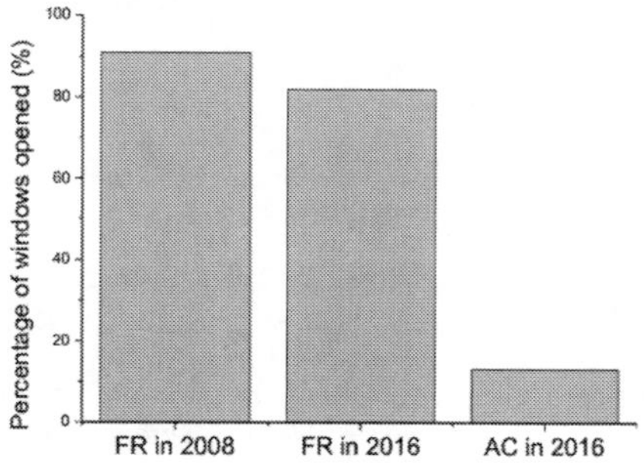

Fig. 8. Proportion of windows opened

4 Conclusions

This paper evaluated occupants' adaptive behaviours and wellbeing in a hot climate before and after the availability of ACs. The following advances and problems were identified:

The most preferred behaviours were using the AC after it became available, followed by using the fan and adjusting the size of the window/door opening. Before that, occupants demonstrated more variety in their behaviours such as drinking something cold, showering, adjusting clothes, sprinkling water on the floor, and changing activities.

The PAU in 2016 and PFU in 2008 demonstrated a logistical fit with the outside temperature. The PFU in 2016 remained at a low level, and it did not change much with the outside temperature.

The main reasons for being unsatisfied with the environment were being too warm, feeling stuffy, perceived bad air quality, and high humidity (>20%). After ACs were installed, occupants' environmental satisfaction vote was higher, and the TSV remained at slightly cool.

When ACs were used, a higher proportion of occupants felt stuffy (54%) and experienced draughts (38%), while a lower proportion of windows were opened (13%); moreover, when ACs were used, more occupants reported being absent-minded (OR = 1.53) and experiencing eye discomfort (OR = 3.22) and discomfort in the upper respiratory tract (OR = 2.58). An analysis of the TSV revealed that the ability of thermal adaption weakened after the ACs were installed.

Thus, althogh the indoor comfort is improved by air conditioner, the indoor air quality in air conditioning sapce should pay more attention in the furture .

Acknowledgements

This research is supported by the Fundamental Research Funds for the Central Universities (Grant No. 2018CDYJSY0055, 2018CDJDCH0015), and the 111 Project (No.B13041).

References

1. Brager, G.S. and R.J. de Dear, *Thermal adaptation in the built environment: a literature review*. Energy and Buildings, 1998. **27**(1): p. 83-96.
2. Wang, J., et al., *Sick building syndrome among parents of preschool children in relation to home environment in Chongqing, China*. Chinese Science Bulletin, 2013. **58**(34): p. 4267-4276.
3. Li, B.Z. and R.M. Yao, *Building energy efficiency for sustainable development in China: challenges and opportunities*. Building Research and Information, 2012. **40**(4): p. 417-431.
4. Zhai, Y.C., et al., *Human comfort and perceived air quality in warm and humid environments with ceiling fans*. Building and Environment, 2015. **90**: p. 178-185.
5. Schiavon, S., et al., *Thermal comfort, perceived air quality and cognitive performance when personally controlled air movement is used by tropically acclimatized persons*. Indoor Air, 2016.
6. Zhu, Y., et al., *Dynamic thermal environment and thermal comfort*. Indoor air, 2016. **26**(1): p. 125-37.
7. Wang, Z.J., et al., *Human thermal physiological and* psychological *responses under different heating environments*. Journal of Thermal Biology, 2015. **52**: p. 177-186.
8. Liu, J., et al., *Occupants' behavioural adaptation in workplaces with non-central heating and cooling systems*. Applied Thermal Engineering, 2012. **35**: p. 40-54.
9. Liu, H., et al., *Seasonal variation of thermal sensations in residential buildings in the Hot Summer and Cold Winter zone of China*. Energy and Buildings, 2017. **140**: p. 9-18.
10. de Dear, R.J. and G.S. Brager, *Thermal comfort in naturally ventilated buildings: revisions to ASHRAE Standard 55*. Energy and Buildings, 2002. **34**(6): p. 549-561.
11. Yao, R., B. Li, and J. Liu, *A theoretical adaptive model of thermal comfort - Adaptive Predicted Mean Vote (aPMV)*. Building and Environment, 2009. **44**(10): p. 2089-2096.
12. Yang, Y., et al., *A study of adaptive thermal comfort* in *a well-controlled climate chamber*. Applied Thermal Engineering, 2015. **76**: p. 283-291.
13. Nicol, J.F. and M.A. Humphreys, *Adaptive thermal* comfort *and sustainable thermal standards for buildings*. Energy and Buildings, 2002. **34**(6): p. 563-572.
14. Rijal, H.B., et al., *Using results from field surveys to predict the effect of open windows on thermal comfort and energy use in buildings*. Energy And Buildings, 2007. **39**(7): p. 823-836.
15. Haldi, F. and D. Robinson, *On the behaviour and adaptation of office occupants*. Building and Environment, 2008. **43**(12): p. 2163-2177.
16. Andersen, R.V., et al., *Survey of occupant behaviour and* control *of indoor environment in Danish dwellings*. Energy and Buildings, 2009. **41**(1): p. 11-16.
17. Schweiker, M. and M. Shukuya, *Comparison of theoretical and statistical models of air-conditioning-*unit *usage behaviour in a residential setting under Japanese climatic conditions*. Building & Environment, 2009. **44**(10): p. 2137-2149.
18. D'Oca, S., et al., *Effect of thermostat and window* opening *occupant behavior models on energy use in homes*. Building Simulation, 2014. **7**(6): p. 683-694.
19. Kim, J., et al., *Understanding patterns of adaptive* comfort *behaviour in the Sydney mixed-mode residential context*. Energy & Buildings, 2017. **141**: p. 274-283.
20. Li, B., et al., *Climatic Strategies of Indoor Thermal* Environment *for Residential Buildings in Yangtze River Region, China*. Indoor And Built Environment, 2011. **20**(1): p. 101-111.
21. Administration, C.M., *The ground climate data of China :Shapingba meteorological station, http://data.cma.cn* 2016.
22. ASHRAE, *ASHRAE standard 55-2013: thermal environmental conditions for human occupancy*. 2013, ASHRAE Atlanta (USA).
23. EPA, *Search Terms (Contains): sick building syndrome*. 2012.
24. Luo, M.H., et al., *The underlying linkage between personal control and thermal comfort: Psychological or physical effects?* Energy and Buildings, 2016. **111**: p. 56-63.
25. Velt, K.B. and H.A.M. Daanen, *Thermal sensation and thermal comfort in changing environments*. Journal of Building Engineering, 2017. **10**: p. 42-46.
26. Kalmár, F., *Investigation of thermal perceptions of subjects* with *diverse thermal histories in warm indoor environment*. Building and Environment, 2016. **107**: p. 254-262.
27. Nicol, F. and M. Humphreys, *Derivation of the adaptive equations* for *thermal comfort in free-running buildings in European standard EN15251*. Building And Environment, 2010. **45**(1): p. 11-17.

Systems analysis of municipal solid waste management and recycling system in east Africa: benchmarking performance in Kigali city, Rwanda

Telesphore Kabera[1], and *Honorine* Nishimwe[2]

[1]College of Science and Technology, University of Rwanda, P.O. Box 3900, Kigali, Rwanda
[2]Development Bank of Rwanda, P.O. Box 1341, Kigali, Rwanda

Abstract. This study assessed the current status of solid waste management (SWM) in the City of Kigali (CoK), the capital city of Rwanda. This assessment was done using systems analysis methodology of "wasteaware" benchmark indicators for integrated sustainable waste management in cities. This method of assessing helps to assess the Municipal Solid Waste Management (MSWM) in a city according to its physical components and governance features. Data were collected during a 6-month period from May to October 2017, but verification took other four months (Up to February 2018). Data were obtained from official reports, legal documents, and interviews with key personnel and operators and also from official and unofficial dumpsites visits. In CoK, approximately 232,870 tons of MSW is generated per year, MSW collection and transportation is done by private companies but the only dumpsite present in Kigali is fully controlled by the City of Kigali. Residents pay waste collectors according to their social classes and their locations. There is no official recycling system; recycling activities are informally performed by private companies and some wastes, like plastic bottles, are taken to Uganda and Tanzania to be recycled. This study found that the recycling rate was 10% contrarily to the 2% which is recorded by official sources. This study has contributed by filling the gap in literature on waste management for the city of Kigali and it recommends that the Government represented by the CoK should do more in terms of investing in SWM and creating a relationship between private waste collectors and local communities, and the private sector should be mobilized to invest in SWM activities.

1 Introduction

Global Waste Management Outlook (GWMO) defined municipal solid waste as the waste from households and small businesses and institutions [1], however, the GWMO notes that the definition varies widely from country to country: that in Rwanda, the Solid waste is defined as all decayable and non-decayable solid and semi-solid wastes, such as trash, garbage, paper, refuse, rubbish, and ashes [2]. [3] mentioned that, although a lot of work has been done to control municipal solid waste disposal in low-income cities in developing countries, there is still some way to go to phase out uncontrolled disposal. In order to deal with SWM problems in developing countries, methods should be developed that can address the local complexities [4]. Furthermore, an approach known as integrated sustainable waste management (ISWM) allows municipalities or districts to manage MSW at the lowest possible cost [5,6].

1.1 City of Kigali

The CoK is the capital of Rwanda, a landlocked country situated in the central/eastern part of Africa. Rwanda is bordered by Tanzania Burundi to the south. In 2016, the gross national income (GNI) per capita was US$700, which is classified as a lower income economy. The CoK's population represents a tenth of the country's population. In 2012, 1,132,686 people resided in the CoK [7]. CoK has is now a home of around 1.5 million resident. CoK consists of three districts and it serves as the economic hub of the whole country. The CoK is facing critical challenges in solid waste management. The rapid increase in solid waste generation in the city and the improper disposal of the generated solid waste result in adverse impacts on health and environment, however, there is a lack of reliable data to assist decision makers for CoK's solid waste management.to **the east,** Uganda to the north, Democratic Republic of Congo to the west and

1.2 Aim of this study

This study aims at analyzing the current state of MSWM of CoK, Rwanda by applying systems analysis methodology of "wasteaware" benchmark indicators for integrated sustainable waste management in cities [8].

2 Methodology

The methodology adopted during this study is based on the city profiling methodology introduced by [1] and developed for UN-Habitat. The methodology was tested and used in more than fifty countries [9, 10, 11; 3]. This use validates the methodology [4].

In order to document and analyse the CoK overall system, a Sankey-style material flow diagram (MFD) was constructed using STAN2.5 software [12] (Figure 1)

The city profiling methodology is based on the integrated solid waste management (ISWM) concept [3]. Data were mainly collected during a 6-month period from May to October 2017, but verification took other four months (Up to February 2018) due to the complexity of municipal solid waste management in the CoK. Data were obtained from official reports and legal documents related to SWM in Rwanda, interviews with key personnel and operators, and also from official and unofficial visits to dumpsites.

3 Material flow analysis for the combined MSWM and recycling system in Kigali

Figure 1 shows the materials flow diagram (MFD) constructed to visualize the combined solid waste and recycling system in the CoK. The sources and methods used for each estimation in the MFD are summarized in Table 1.

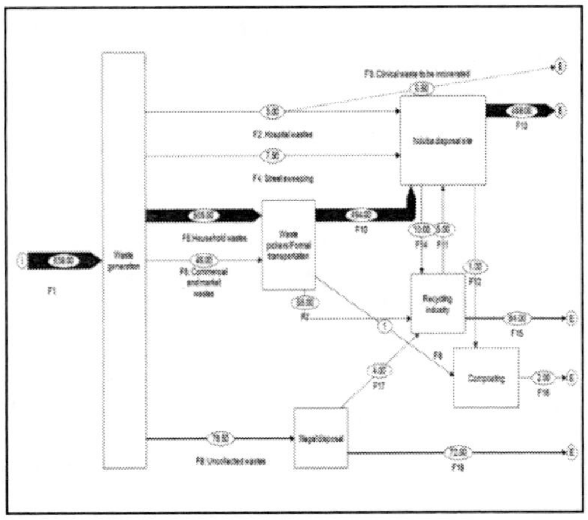

Fig 1. Sankey type material flow diagram (MFD) for MSW in the City of Kigali (the thickness of the flow corresponds to the waste value.)

There is an acute lack of data on waste management and recycling in CoK. There are no weighbridges in daily use, so use had to be made of data on the number of vehicles arriving at the site from each source and on average loads. In 2014, vehicles were weighed at Nduba disposal site and only two types of vehicles were distinguished: one type carries 10 tonnes on average and the other 5 tonnes. Total waste generation is seldom measured. Here it is back-calculated from estimates of the quantities of waste collected and of collection coverage (see Table 1).

Table 1. Basis for calculating the waste flows for Kigali in the MFD

Flows	Description	Calculation	Basis for data
F1	Total Quantity of waste generated in Kigali-City	Official data give collection coverage as 88% F1=(F2+F4+F5+F6)/0.88	Back-calculated from collection coverage and other flows in MFD
F2	Waste from official Health Centers and Hospitals	Calculated based on the average load delivered by the vehicle. 3 vehicles a week each carrying average load of 5 tonnes. 15 tonnes over 6 days, 2.5 tpd collected.	Vehicle numbers and load size provided by the person in charge of Nduba disposal site
F3	Hazardous clinical waste (Incineration is used as treatment process)	Assumes 0.25 kg per bed per day (cf WHO guideline 0.2 (Haifete et al., 2016)). There are 2000 beds in the 5 main hospitals.	Estimate by authors, informed by WHO data
F4	Waste from street sweeping/ street litter bins	9 vehicles per week each carries 5 tonnes. 45 tonnes over 6 days = 7.5 tpd	Vehicle numbers arriving at Nduba disposal site
F5	Waste collected from door to door by the contracted private companies.	30 vehicles per day carrying 10 tonnes + 41 vehicles per day carrying each 5 tonnes.	Vehicle numbers and sizes arriving at Nduba disposal site
F6	Waste collected from commercial areas/markets	9 vehicles per day carrying each 5 tonnes	Vehicle numbers arriving at Nduba disposal site
F7	Wastes collected by the	Estimated at 10% of flows F4 and F5	Estimate by authors based on observation and

 https://doi.org/10.1051/e3sconf/20198003004

	private companies and separated by their employees for recycling	F7 = 0.1*(F5+F6) = 0.1*550 = 55 tpd	interviews
F8	Organic wastes collected directly from the source and transported to compost plant	Very small flow. Estimated at 1 tpd	Estimate by authors based on observation and interviews
F9	Uncollected wastes	Waste collection coverage equals 88%. So, 12% is not collected. Then F9 = 12% of F1	Collection coverage calculated from official data
F10	Waste delivered to the dumpsite site from households, commercial areas and markets	Waste collected less the quantity separated for recycling F10= F5 + F6 – F7	Calculated from the MFD
F11	Reject materials from recycling sent to Nduba disposal site	Estimate	Estimate by authors based on observation and interviews
F12	Organic waste sorted by waste pickers from Nduba disposal site, taken to composting plant	Estimate	Estimate by authors based on observation and interviews
F13	Dumpsite materials	F13= (F2-F3 + F4 +F10 + F11) – (F14 + F12)	Calculated from the MFD
F14	Waste sorted for recycling by pickers at the dumpsite site.	Estimate. Mainly plastic bottles, exported to Uganda and China	Estimate by authors based on observation and interviews
F15	Recycled materials	F15 = (F7 + F17+ F14) – F11	Calculated from the MFD
F16	Fertilizer from Compost plant.	F16 = F8 + F12. Quantities very small, so no attempt to correct for process losses.	Calculated from the MFD.
F17	Waste sorted for recycling by waste pickers on illegal dump-sites	Estimate	Estimate by authors based on observation and interviews
F18	Waste left in illegal dumpsites after waste picking	F18 = F9 – F17	Calculated from the MFD

4 Systems analysis of the MSWM and recycling in Kigali

4.1 Waste generation and composition

According to the most recent Rwanda Population and Housing Census (RPHC4), the population of Kigali in 2012 was 1,132,686 [7]. The growth rate since 2002 has been 2.6% per annum [7]; using this to extrapolate forward to 2017, the population used here is 1,574,000.

The MFD (Figure 1) estimated waste generation at 638 tonnes per day, or 233,000 tonnes per year. This corresponds to a per capita MSW generation of of 0.41 kg/day, or 148 kg/year. No specific data could be located for waste composition in Kigali or Rwanda. We have used data from a 2012 planning study, which did not measure waste composition directly but instead used figures 'typical for a low income country': organics 70% by weight; paper 6%; plastic 5%; and metals 3% [13]. No data are available for waste density, moisture content or calorific value.

4.2 Waste collection

The official report from the Office of the Auditor General of State Finances, Rwanda mentioned that only 2% of solid waste is recycled, but our investigation has shown the recycling rate to be 10% (55 tonnes per day). The 10% rate represents separated wastes collected by private companies, from households and commercial areas for recycling. The basic system is that householders store their waste in individual containers, which are collected door-to-door once a week; while commercial areas and markets use communal containers (sizes of containers are 60, 120, 240, 660, 1100 liters)or walled collection bays (sizes of walled collection bays is

7x4x2.5 m^3), from which waste is collected daily. The system works reasonably well, with three medium/high compliance scores of 15/20 and three of 10/20, giving an overall score of 63%, i.e. a medium/high assessment for the quality of waste collection and street cleaning.

4.3 Waste treatment and disposal

There is only one disposal site in Kigali city, called Nduba after its location in Gasabo district. The site started operations in 2012 and is run by the City. Nduba scores 5 or 10/20 on the three 'quality criteria', which categorises it as semi-controlled, below the threshold set for controlled disposal. As there is no controlled disposal facility in use in Kigali, the Wasteaware 'controlled disposal' indicator is 0%. The quality indicator is rated Medium/Low (30% score)

4.4 Resource Management

The overall assessment is Low/Medium, with a score of 33% In the CoK, awareness support for waste separation at the source (householders and offices) is very high but people do not separate waste at the source. In the past, communities used to sort and separate the waste at the source, but due to irregular collections they lost trust in the waste collectors. Regular collection of the sorted waste increases trust from service users [14].

4.5 Governance

4.5.1 User Inclusivity (UI)

The benchmark indicator for user inclusivity assesses citizens' and other waste generators' inclusion in the waste management system. The overall assessment is medium/high, with a score of 71%. Access of users to services is generally good. Numerous systems are in place for citizens to make their voices heard, including environmental committees at 4 levels of local government from the 'cell' upwards, which then pass on their concerns either at formal monthly meetings with CoK officials, or through the monthly 'Umuganda' community service day. The CoK carries out proactive public awareness and education, but is constrained by lack of resources. The evidence for people actually 'doing the right thing' is also strong.

4.5.2 Provide Inclusivity (PI)

The score here was found to be 55 which represents medium as an overall assessment. Generally In the CoK, private companies in charge of SW collection provide high quality *service.*

4.5.3 Financial Sustainability

Financial sustainability benchmark indicators assess how the city's SWM service is financially sustainable. Total score is 50% which corresponds to medium as an overall assessment. According to one in charge of SWM under the CoK, the budget allocated to SWM is not sufficient.

4.5.4 Sound institutions and proactive policies: National SWM Framework

These benchmark indicators assess the adequacy of the national solid waste management framework; this assessment also includes the degree of implementation. The overall assessment is medium, with a score of 58%. In Rwanda there are a lot of legislations and regulations on SWM but their enforcement is weak (for example there is prohibition of manufacturing, importation, use and sale of polythene bags in Rwanda but still some people use plastic bags illegally entered into Rwanda).

4.5.5 Sound institutions and proactive policies: Local Institutional Coherence

These assess the institutional strength and coherence of the city's solid waste management functions. The overall assessment is medium, with a score of 50%.

The level of collaboration between governmental institutions and private companies is sufficient. Data availability is limited at Nduba disposal site (no weighbridge, nothing about waste generation etc...). There is an urgent need to improve data collection, reporting methods and monitoring methods.

5 Discussion

This study has assessed the CoK's MSWM. People in the CoK generate wastes equal to 232,870 tonnes/year and waste generation per capita is equal to approximately half (405g/capita/day) of a higher waste generating city (above 1 kg/capita/day). This explains that the CoK belongs to lower generating cities and then there is no urgency of waste reduction, apart from the issue to take critically the diversion from treatment and disposal to recycling.

For public health-waste collection category, CoK showed a strong performance. Regarding, the environmental control-waste category, CoK performes well, however, there is still more to be done, CoK should make efforts to improve environmental protection measures, a waste treatment plant should be built at Nduba disposal site, leachate should be controlled and workers at Nduba disposal site should be protected. The score for recycling rate was low and this shows that Rwanda needs a strong improvement in order to increase the recycling rate. For CoK, 3R (reduce, reuse, and recycle) quality indicators scored 33%. This score is classified as low-medium performance and in order to improve 3R quality indicators the CoK should urgently start concentrating its efforts on recycling activities.

Table 2. Wasteaware benchmark indicator results for Kigali city

No	Category	Indicator		
	City		Kigali	
	Country		Rwanda	

Background information on the city				
	Country income level	World Bank income category	Low	
B1		GNI per capita	$700	
B2	Population	Total population of the city	1,574,434	
B3	Waste generation	MSW generation (tones per year)	232,870	
Key waste-related data				
W1	Waste per capita	MSW per capita (kg per year)	148	
W2		Waste composition		
W2.1	Organic	Organic (foods and green waste)	70%	
W2.2	Paper	Paper and card	6%	
W2.3	Plastic	Plastic	5%	
W2.4	Metals	Metals	3%	
Physical components				
1.1	Public health-waste collection	Waste collection coverage	88% (M)	
1.2		Waste captured by the system	88% (M)	
1C		Quality of waste collection service	63% (M/H)	
2	Environmental control-waste	Controlled treatment and disposal	79 (M)	
2E		Degree of environmental protection in waste treatment and disposal	30 (L/M)	
3	Resource management-	Recycling rate	10% (L/M)	
3R	reduce, reuse and recycle	Quality of 3Rs- Reduce, reuse, recycle-provision	33% (L/M	
Governance factors				
4U	Inclusivity	User inclusivity	71% (M/H)	
4P		Provider inclusivity	55% (M)	
5F	Financial sustainability	Financial sustainability	50% (M)	
6N	Sound institutions	Adequacy of national SWM framework	58% (M)	
6L	Proactive policies	Local institutional policies	50% (M)	

For user inclusivity qualitative indicator (4U), the score was good (71%) and only needed improvement in the consideration of local communities' contributions to decisions. Regarding the provider inclusivity qualitative indicator (4P), the score was 55%. The most needed improvements are the representation of the private sector and the integration of the informal sector into the formal WMS because they have knowledge on waste composition and recyclables [4].

The score for financial sustainability was found to be 50% and below for four cities out of five except Maputo which scored 67%, this shows that the financial sustainability of the SWM system is a major challenge in the CoK and in other cities of developing countries [4]. This financial sustainability challenge can be overcome by sensitizing the private sector to invest in solid waste management and sensitize local communities to work closely with the private sector. In the eyes of local communities, private companies in charge of waste collection are seen as companies which are only after money. A lot of effort should be made to build a strong relationship between the private sector and local communities. The city has good policies, strategies, legislation and regulations but the problem is the implementation. For example, CoK has ten sets of good strategies but none of them has been implemented so far (6N). Government institutions and the CoK (6N and 6L) in particular have to make sure that all those good strategies and policies are fully implemented.

6 Conclusions and recommendations

This paper has presented the SWM profile of the CoK. The country has adopted a lot of strategies to protect the environment such as the legal ban on use of non-biodegradable plastic bags. Despite making a lot of effort towards SWM improvements, there are still considerable improvements to be made. The study showed that data such as moisture content and solid waste density are not known. In order to raise the standard of solid waste management in the CoK, this lack of data problem should be addressed.

In order to conduct this study a systems analysis methodology of "wasteaware" benchmark indicators for integrated sustainable waste management in cities was applied. For allowing a visualization of the MWMS in the CoK a MFD diagram was provided to accompany the "wasteaware" benchmark indicators.

This study estimated the MSW generation rate in the CoK at 0.405 kg/capita/day with a total 638 tonnes per day of waste. In order to estimate the waste generation per capita, the number of vehicles bringing waste at Nduba disposal site of the City of Kigali were used. The major information found during this study is the recycling rate of 10%, which is contrary to the recycling rate from official sources (2%). In terms of sound institutions and proactive policies, both national SWM and local institutional coherence are rated medium (the scores were 58% and 50% respectively). The only needed improvement here is to monitor the implementation of suitable proposed policies/strategies.

Government represented by the CoK should do more in terms of creating a relationship between private waste collectors and local communities so that local communities stop looking at these private companies as only money-oriented companies. The private sector should be sensitized to invest in SWM activities. The CoK and private waste collecting companies should also improve working conditions for their workers.

Reference

1. Wilson C David C , Rodic Ljiljana, Cowing J Michael, Velis A Costas, Whiteman D Andrew, Scheinberg Anne, Vilches Recaredo, Masterson Darragh, Stretz Joachim and Oelz Barbara (2014) 'Wasteaware' Benchmark Indicators for Integrated Sustainable Waste Management in Cities. Supporting Information For paper published in Waste Management – doi: 10.1016/j.wasman.2014.10.006. Available at http://wasteaware.org/wp-content/uploads/2015/11/Wilson_et_al_Supplementary_information_Wasteaware_ISWM_Benchmark_Indicators_User_Manual_Online.pdf

2. MININFRA (2016) National Sanitation Policy, Kigali-Rwanda. Available at: http://www.rura.rw/uploads/media/NATIONAL_SANITATION_POLICY__DECEMBER_2016.pdf.
3. Wilson C David , Rodic Ljiljana, Scheinberg Anne , Velis A Costas and Alabaster Graham (2012) Comparative analysis of solid waste management in 20 cities, Waste Management & Research 30(3) 237–254.
4. Byamba Bolorchimeg and Ishikawa Mamoru (2017) Municipal Solid Waste Management in Ulaanbaatar, Mongolia: Systems Analysis, Sustainability 2017, 9, 896; doi:10.3390/su9060896.
5. Anschütz, J., IJgosse, J. and Scheinberg, A. (2004) Putting ISWM to Practice. Using the ISWM Assessment Methodology, ISWM Methodology as applied in the UWEP Plus Programme (2001-2003), WASTE, Gouda, The Netherlands.
6. Van de Klundert, A. and Anschütz, J. (2001) Integrated Sustainable Waste Management – the concept. Tools for decision-makers, Experiences from the Urban Waste Expertise Programme (1995-2001), WASTE, Gouda, the Netherlands.
7. National Institute of Statistics of Rwanda (2012) The fourth Population and Housing Census in Rwanda. Available at: http://microdata.statistics.gov.rw/index.php/catalog/65.
8. UNEP and ISWA (2015). *Global Waste* Management *Outlook*. Editor-in-Chief: Wilson D C Authors: Wilson D C, Rodic L, Modak P, Soos R, Carpintero A, Velis C A, Iyer M and Simonett O. Prepared for UNEP and ISWA. Osaka: UNEP. Available at: https://www.researchgate.net/publication/283085861_Global_Waste_Management_Outlook_United_Nations_Environment_Programme_UNEP_and_International_Solid_Waste_Association_ISWA (Accessed 13.07.2018).
9. Al Sabbagh MK, Velis CA, Wilson DC and Cheeseman CR (2012) Resource management performance in Bahrain: a systematic analysis of municipal waste management, secondary material flows and organizational aspects. Waste Management & Research 30 (8):813-824.
10. Masood, M.; Barlow, C.Y.;Wilson, D.C. (2014) An assessment of the current municipal solid waste management system in Lahore, Pakistan. Waste Manag. Res. 32: 834–847.
11. Sim, N.M.; Wilson, D.C.; Velis, C.A.; Smith, S.R. (2013) Waste management and recycling in the former Soviet Union: The City of Bishkek, Kyrgyz Republic (Kyrgyzstan). Waste Manag. Res. 31: 106–125.
12. Cencic O and Rechberger H (2008) Material flow analysis with software STAN. Environmental Engineering and Management Journal 18: 3–7.
13. BURGEAP (2012) *Solid waste management plan final report*. Report to City of Kigali, funded by Belgian Government. Kigali, Rwanda.
14. Kodwo Miezah , Kwasi Obiri-Danso , Zsófia Kádár, Bernard Fei-Baffoe, Moses Y. Mensah(2015) Municipal solid waste characterization and quantification as a measure towards effective waste management in Ghana. Waste Management 46 (2015) 15–27.
15. Kawai Kosuke; Tasaki Tomohiro (2016) Revisiting estimates of municipal solid waste generation per capita and their reliability, J Mater Cycles Waste Manag 18:1–13 DOI 10.1007/s10163-015-0355-1.

E3S Web of Conferences 80, 03005 (2019) https://doi.org/10.1051/e3sconf/20198003005
REEE 2018

Investigation on heat stress of construction workers in summer in Chongqing, China

Longqian Li[1,2], *Hong* Liu[1,2], *Lu* Chen[1,2], *Jiaze* He[1,2]

[1]Joint International Research Laboratory of Green Buildings and Built Environments (Ministry of Education), Chongqing University, Chongqing 400045, China
[2] National Centre for International Research of Low-carbon and Green Buildings (Ministry of Science and Technology), Chongqing University, Chongqing 400045, China

Abstract. The heat stress of workers at construction sites in Chongqing, China, was studied in this paper. The investigation included measurement of environmental parameters, measurement of physiological parameters, questionnaire survey and tracking record of their self-regulation of the behaviour. Scaffolder, carpenter, porter, plasterer and handymen were taken as the main research subjects. Five subjects were selected daily to conduct a questionnaire survey and to measure their physiological parameters, 60 construction workers were investigated. The results show that in sizzling summer, the average effective Wet Bulb Globe Temperature (WBGTeff) is 35.8 ° C. Under these circumstances, workers are in a poor working environment based on some Chinese and international standards. During the experiment, the researchers recorded the work behaviour and self-regulation of workers. It is found that the average heart rate of workers with different job is significantly different, and the average metabolic rate of handymen is significantly lower than that of piling rig workers. Different type of workers should be protected by different protection rules to avoid suffering heat injury.

1 Introduction

With the economic development and social progress, people began to pay more and more attention to the occupational hazards in all kinds of work. In recent years, Wet Bulb Globe Temperature (WBGT) in various parts of the world has hit record highs in summer. In hot temperature environment, engaged in a certain intensity of manual labor, if not fully protected, easy to accumulate heat due to the body, causing the core temperature, leading to normal functioning of internal organs decline in body and water and salt imbalances and other issues. When the heat storage effect exceeds the physiological range of human body temperature regulation, it may lead to several types of heat diseases. In the summer of 2018, several cases of heat diseases have taken place in Chongqing City, Shaanxi Province, Jiangsu Province and Macao Special Administrative Region in China, causing dozens of people died, including outdoor workers such as construction workers and sanitation workers.

Although, as technology advances, labor in some extremely harsh environments has been replaced by various machines or equipment, much remains to be done by manual work, such as long-term outdoor work in the construction workers, sanitation workers, postmen. When the weather is hot outdoors in summer, these types of work are more susceptible to heat load.

Compared with other industries, the construction industry is more susceptible to high-temperature environments under summer high-intensity solar radiation[1]. At present, Chinese construction industry lacks standards for protecting workers in high-temperature environments, and its management level lags behind that of developed countries. The protection of workers is limited to such decrees as the 《Notice on the Administration of Measures for the Administration of Heatstroke Prevention and Cooling Measures》. Laws, regulations, standards and protection systems to protect construction workers from heat damage need to be established. The purpose of this study is to investigate the working environment of the construction site and the health status of workers, and to provide reference materials for labor protection of workers in high temperature environments.

2 Research methods

2.1 Site investigation

This research mainly adopts the method of site investigation. A construction site in Banan District, Chongqing, China, was selected to conduct this research,during July to August in 2018. More than 500 workers work at the same time every day at the construction site. The climate of is hot in summer and cold in winter. The temperature in summer is mostly between 27.0 ℃ and 38.0 ℃, and the maximum temperature can reach 43.8 ℃. The hottest 15 days with temperatures above 35.0 ℃ were selected for research. The construction site of the job is generally divided into scaffolding workers, handymen , carpentry, porters, steel workers, bricklayer, etc, which contains the upper extremity exercise, lower extremity exercise and comprehensive exercise. In terms of schedule, during the hot summer, the managers arranged for workers to work from 6 am to 11 am and from 4 pm to 8 pm , avoiding the hottest hours.

The field studies included measurements of environmental and physiological parameters , questionnaires , and observations of workers' voluntary behavioral adjustments. Sixty workers with different work were randomly selected as subjects,five workers were tested daily, and each worker was tested for a half day for a total of fifteen days.

2.2 Measurement of environmental parameters

According to the site environment, WBGTeff was chosen to evaluate the workers' environment. WBGT (Wet Bulb Globe Temperature) is the most widely used direct indicator of thermal stress today, it is to deal with the event of heat casualties of US naval training [2] The index combines three environmental parameters of air temperature (Ta), natural wet-bulb temperature (Tnw) and black globe temperature (Tg), considering four basic environmental parameters such as dry bulb temperature, wet bulb temperature, radiation temperature and air velocity, can be applied to different environments. According to ISO 7243: 2017[3-6] ,WBGT is calculated as

Without solar load

$$WBGT=0.7t_{nw}+0.3t_g \quad (1)$$

With solar load

$$WBGT=0.7t_{nw}+0.2t_g+0.1t_a \quad (2)$$

These formulas are applicable to cotton work clothes (thermal resistance of clothing is 0.6 clo and impermeability index is i_m=0.38). different clothing, especially with a different evaporative resistance, is likely to have a different effect on the heat stress level. For clothing materials and configurations different from standard work clothing, clothing adjustment values (CAVs) in WBGT temperature units are provided. The CAV is added to the measured WBGT to produce an effective WBGT($WBGT_{eff}$) that represents an estimate of the heat stress provided by the actual clothing worn as an equivalent environment.

$$WBGT_{eff}=WBGT+CAV \quad (3)$$

At the construction site, WBGT, air temperature, black globe temperature and air velocity were measured. Portable WBGT Thermal Index Device, model no. JTR10A (precision ±0.5℃) is used to measure WBGT, air temperature and globe temperature. KIMO handheld hot wire anemometer, model no. VT100 (precision ±0.1m/s) is used to measure air velocity.

2.3 Measurement of physiological parameters

The purpose of high-temperature environmental safety management is to monitor and control the impact of environmental heat load on human physiological heat stress and avoid the generation of heat diseases. The physiological response of workers in high temperature environments is extremely complex , which will also be directly affected by individual differences in workers. During work, sweat will take away salt and water in the body. Copious amounts of sweat cause a lot of body fluid loss, blood concentration increases, body temperature regulation ability decreases, heart rate increases, work efficiency decreases, and workers' heart rate, blood pressure and body temperature increase accordingly. Allow workers to rest when they reach the threshold they can withstand, and work again after a while when the body resumes. The main physiological parameters are heart rate, blood pressure, ear temperature and skin temperature.

The metabolic rate determines the degree of comfort or fatigue exposed to the thermal environment. Particularly in hot weather, a large amount of metabolic heat generated by muscle activity exacerbates heat stress, and a large amount of heat needs to be dissipated in a sweat evaporation manner. Heart rate, as an alternative to measuring energy metabolism, suggests that there is a correlation between heart rate and energy metabolism during moderate-intensity physical activity[7-8]. At the same time, according to ISO 8996: 2004, in the case of dynamic work using major muscle groups, with only a small amount of static muscular load and in the absence of thermal strain and mental loads, the metabolic rate may be estimated by measuring the heart rate while working. The relationship between heart rate and metabolic rate can be written as:

$$HR=HR0+RM\times(M-M0) \quad (4)$$

M : metabolic rate(W/m^2)

M_0: metabolic rate at rest (W/ m^2)

RM : the increase in heart rate per unit of metabolic rate

HR_0 : the heart rate at rest, under neutral thermal conditions. (bpm)

$$RM=(HR_{max}-HR_0)/(MWC-M_0) \quad (5)$$

MWC :The maximum working capacity （W/ m^2）

$$Men:MWC=(41.7-0.22A)P^{0.666} \quad (6)$$

$$Women:MWC=(35.0-0.22A)P^{0.666} \quad (7)$$

$$HR_{max}=205-0.62A \quad (8)$$

A :age(in year)

P : weight (in kg)

Without affecting the normal work of workers under the premise, Polar heart rate sensor, model no.A300 (precision ±1%) is used to measure heart rate. Also, Barun IRT6520 thermoscan ear thermometer, DS1922T temperature logger and Omron U30 electric blood pressure monitor is used to measure ear temperature, skin temperature and blood pressure respectively.

2.4 Questionnaire

The collection of basic personal information includes height, weight, body temperature, average blood pressure, medication, related medical history, work clothes, feelings of the working environment, expectations, education level, time schedule, adjustments etc, which to some extent reflect the physical condition of the worker. Recording subjects' working hours, smoking frequency, drinking habits and leisure activities can reflect whether workers are overworked and their lifestyle is healthy. The follow-up survey mainly records the type of work of the tested workers, the situation of the workplace, the type of drinks, the frequency of drinking water, and adjustment methods

2.5 Experiment process

Firstly, a WBGT monitor was placed at the worker's workplace to record environmental parameters. The researchers then helped the workers fill out the questionnaire and gave the workers a heart rate belt, measuring their blood pressure and ear temperature. Subsequently, the experiment began, workers started working, their heart rate was continuously monitored, and the recording interval was 1 s. WBGT, air temperature, wet bulb temperature, black bulb temperature and wind speed are recorded every half hour. After the workers started working, the researchers observed and recorded their work behaviors and adjustment behaviors. After the work, the workers moved to a cool place to rest, and the researchers measured their heart rate, blood pressure and ear temperature.

3 RESULTS

In the survey, there were 16 carpenter, 12 scaffolding workers, 8 porters, 8 handymen and 16 other workers, for a total of 60 people. The youngest is 27 years old, who is the only worker aged 20 to 30. The oldest is 65 years old. There are 36 elderly people aged 50 to 70, accounting for 60% of the total number of surveys, showing a clear trend of aging. There are 17 workers with a working age of less than 5 years, 30 with a working age of 5 to 20 years, and 13 with a working age of more than 20 years. The workers with the largest working age are about 50 years who is 62 years old.

In the survey, 44 people said that they had experienced discomfort such as dizziness, chest tightness and tinnitus during the high temperature working hours in the afternoon. It is found that workers working in certain places are not conducive to relieving heat stress. For example, when workers are scaffolding, due to space constraints, workers with aloft work obviously drink less water than workers who work on the ground. Workers who plaster in poorly lit rooms. Due to workers eager to finish their work and get out of the dark environment. also drink water at a lower frequency. The other hand, Carbon dioxide in cigarettes reduces the amount of oxygen in the blood, while drinking affects the ability to respond and increases the workload of the liver, both of which cause the body to experience reduced function.however, the survey found that 38 workers had smoking behavior in addition to drinking water and chatting during work breaks. They think smoking helps to relieve their thermal stress.

3.1 $WBGT_{eff}$

Table 1 shows the values of average environmental parameters of workers during continuous working hours during the experiment period. According to the survey, workers wear clothes made of cotton or synthetic fibers, corresponding to work clothes and cloth coveralls in ISO 7243: 2017 annex F,both of which have zero adjustment values for clothing. Even avoiding the hot noon period, the average WBGTeff per day was as high as 29.6~40.2°C and the average fifteen-day air temperature was 38.4°C, average wet bulb globe temperature was 37.7°C. According to ISO 7243 and ISO8996, [4,6]The metabolic rate grades of several workers investigated were mainly moderate metabolic rate (300W) and high metabolic rate (415W). Considering that subjects are long-term construction workers,workers exercise moderate metabolic rates, at which time the values of WBGT reference limit for persons acclimatized to heat is 28.0 ° C. According to some of Chinese recommended standards, such as GB/T 17244, GB/T 4200 and GBZT 299.3, workers in a poor thermal environment are likely to cause health hazards and heat damage to workers. Require managers to take occupational health guardianship and sunstroke protection measures, heat stress monitoring and reduce the heat stress response and the unit of time ratio by adjusting the high-temperature work-rest system.

Table 1. environment parameters of the workplace

Day	Black globe temperature (°C)	Air temperature (°C)	Wet bulb temperature (°C)	$WBGT_{eff}$ (°C)
1	37.7	40.3	39.8	38.4
2	39.0	43.7	41.3	40.2
3	36.7	38.7	38.4	37.3
4	35.0	37.1	36.9	35.6
5	34.2	36.3	36.0	34.8
6	35.8	38.3	37.8	36.5
7	35.0	37.4	36.9	35.7
8	35.9	38.4	38.1	36.6
9	37.0	39.5	38.9	37.7
10	28.0	33.5	32.9	29.6

https://doi.org/10.1051/e3sconf/20198003005

11	30.4	40.5	38.7	33.3
Average	35.0	38.5	37.8	35. 8

3.2 Heart rate and metabolic rate

Table 2 shows the average heart rate of workers in different types of work and the highest and lowest average heart rate and metabolic rate per minute in the experiment. The work of the scaffolding workers is to install, disassemble and transport the steel pipe. The working environment of the outer Scaffolding workers is poor, and the average metabolic rate, the average heart rate during work and rest are obviously higher than those of the indoor scaffolding work; carpenter cuts wooden boards and builds concrete molds. working in shaded areas, the average heart rate difference between work and rest is 21.45bpm. The work of the handyman is mainly for the auxiliary work of sorting and recycling steel bars、 wood molds and chiseling grooves. Working in shaded areas, the average heart rate difference between work and rest is 10.65bpm. Its labor intensity is low; The porters mainly carry heavy materials such as I-beams and sand, and most of them work in places where they are completely shaded, The labor intensity is large, but their working time is intermittent. The instantaneous heart rate of moving heavy objects increases greatly. They are rest while waiting for the tower crane to transport heavy loads. Its working environment is better than the hot outdoor, the average heart rate difference between work and rest is 19.16bpm. The plasterer is mainly for upper limb movement and works indoors , the average heart rate difference between work and rest is 14.00bpm. The work of the plumber is mainly for digging trenches, landfill pipelines and installing pipelines.

Working indoors, the average heart rate difference between work and rest is 16.76bpm; The piling rig worker operates the drill bit-pile to piling the well. , their needs to hold the drill bit by hand for a long time. It has a high labor intensity and works in an outdoor hot environment. The average heart rate during work is 126.60bpm, and the average heart rate at rest is 26.03bpm, the average heart rate of work time is higher than that of rest about 26.03bpm ; According to ISO8996, this type of work is classified as a high metabolic rate level（200~260W/m^2）[4]; The steel workers work in the outdoor steel shed. The steel bars transported to the construction site by the factory are made into the shape required by the construction process. Since the construction organizer only arranged a small number of steel workers to work on site during the investigation, in order not to affect the progress of the project, only one steel worker participated in the survey. His resting average heart rate was 87.70bpm, the average heart rate during work was 116.72bpm, and the average heart rate changed 29.02 bpm. Metabolic rate of this worker reaches 213w/m^2. The average metabolic rate of all workers was 151.21 W/m^2.

3.3 The average metabolic rate with different types of work

The increase in heart rate mainly is due to heat stress and dynamic muscular load, under neutral thermal conditions.As is shown in Figure 1-4,the worker's resting heart rate and heart rate at work as well as age and weight information into formula, draw the average metabolic rate of all workers . compared with the average metabolic rate with different types of work, it could find that steel worker and piling rig worker is the hardest work in this survey,because they have to work in a environment that is directly exposed to sun,In addition The average metabolic rate of the handyman is the smallest, consistent with the results observed by the researchers, their labor intensity is relatively low, and it does not work with direct sunlight. The horizontal line in the figure is the average metabolic rate of all workers.

Table 2. Average heart rate of workers in different type of work and metabolic rate

JOBS	Resting average HR (bpm)	Working average HR (bpm)	Working-resting (bpm)	Minimum average HR (bpm)	Maximum average HR (bpm)	Average Metabolic (W/m^2)
Outdoor scaffolding workers	82.74	103.35	20.61	74.29	128.86	160.93
scaffolding workers inside building	79.49	96.42	16.94	69.00	122.40	128.97
carpenter	85.32	106.77	21.45	73.38	135.31	159.36
porters	82.38	101.55	19.16	74.50	131.50	139.54
handymen	77.21	87.86	10.65	67.25	109.00	98.13
plumber	84.69	101.44	16.76	74.40	123.00	132.65
plasterer	87.99	101.98	14.00	74.00	141.75	127.16
piling rig worker	100.57	126.60	26.03	90.50	150.50	201.10

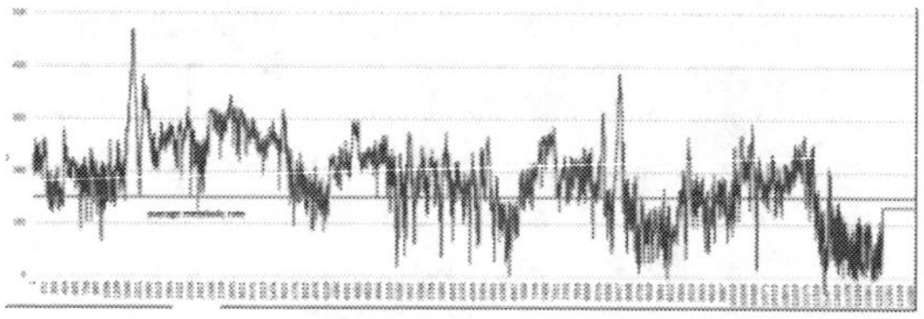

Figure 1. Average metabolic rate of steel workers

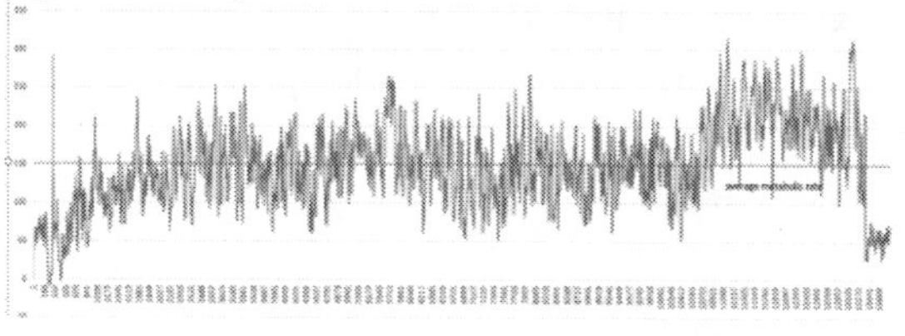

Figure 2. Average metabolic rate of carpenter

Figure 3. Average metabolic rate of outdoor scaffolding workers

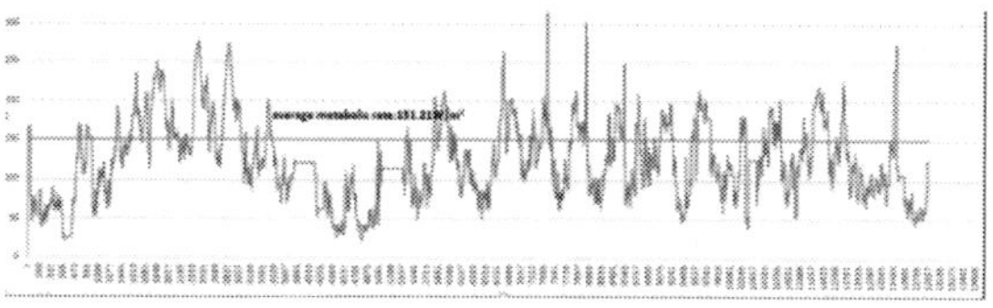

Figure 4. Average metabolic rate of porters

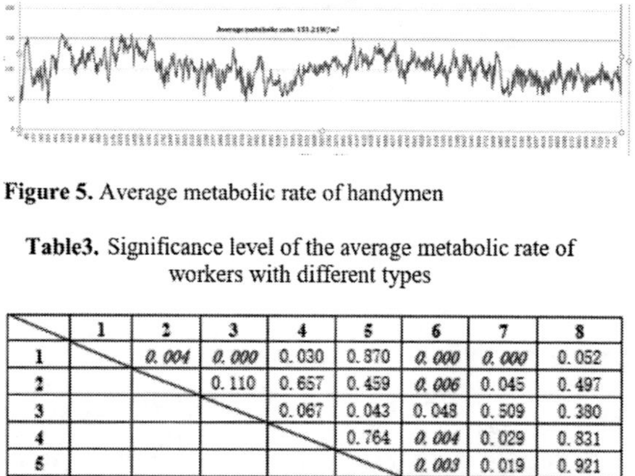

Figure 5. Average metabolic rate of handymen

Table3. Significance level of the average metabolic rate of workers with different types

	1	2	3	4	5	6	7	8
1		*0.004*	*0.000*	0.030	0.870	*0.000*	*0.000*	0.052
2			0.110	0.657	0.459	*0.006*	0.045	0.497
3				0.067	0.043	0.048	0.509	0.380
4					0.764	*0.004*	0.029	0.831
5						*0.003*	0.019	0.921
6							0.122	*0.003*
7								0.016
8								

1- handymen ; 2-porters ; 3-carpenter ; 4-plumber ; 5-plasterer ; 6- piling rig worker ; 7-Outdoor scaffolding workers; 8- scaffolding workers inside building

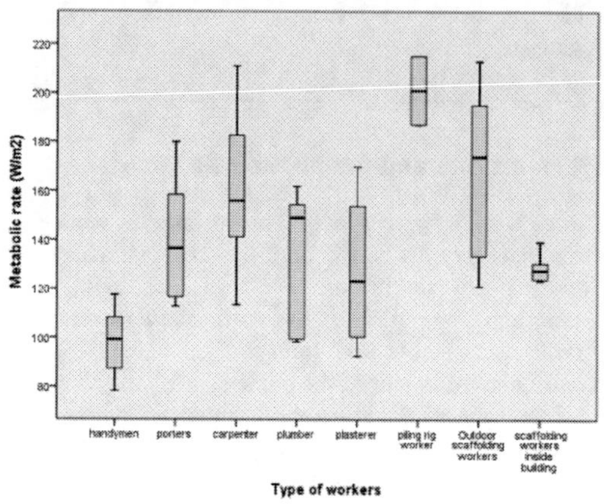

Figure 6. Boxplot of the average metabolic rate of workers with different types

It can be found from Table 3 that there is a significant difference in the average metabolic rate between the handyman and the porter, piling rig worker, and Outdoor scaffolding workers ($P<0.005$). Due to the low labor intensity and the fact that most of the handymen are not working in direct sunlight,the average metabolic rate of the handyman is significantly lower than that of other workers. On the contrary, the average metabolic rate of the digging well was significantly higher than that of other workers ($P<0.01$)，which due to their high labor intensity and the direct sunlight in their workplace.

4 Discussion and Conclusions

In 2018, Chongqing issued "high temperature red warning" signals many times, and temperatures in many districts and counties reached 40.0°C or above.Workers are in a very poor thermal environment, both in China and in international standards, it is very likely to cause the worker's health hazard and generate heat damage. although there are significant individual differences in metabolic rate among workers，it is found that the trend of heart rate and metabolic rate of workers in the same type of work is similar. and that the average metabolic rate of workers with different jobs is obviously distinctive. Different types of work may face different risks of thermal damage. Such as steel workers, piling rig workers, and scaffolding workers who work outside and who have relatively high labor intensity should receive more comprehensive thermal protection measures. It is recommended to wear relevant physiological parameter monitoring instruments to protect workers' safety in case of thermal disease. Adjusting of workers may be hampered by certain restrictions on a construction site. On the other hand ,such behaviours as working on a scaffolding or plastering in a poorly lit space would affect the worker's frequency of drinking or resting. The impact of workplace space constraints on self-regulation of workers is also a potential research direction.

Acknowledgements

This research is supported by The UK-China collaborative research project "Low carbon climate-responsive Heating and Cooling of Cities (LoHCool)" supported by the National Natural Science Foundation of China (NSFC Grant No. 51561135002) and UK Engineering and Physical Sciences Research Council (EPSRC Grant No. EP/N009797/1), and the 111 Project (No.B13041).

References

1. Chen D X, Bai S G, Zeng Z S. 1992. Classification of physical strength of construction workers. Chinese Journal of Industrial Hygiene and Occupational Diseases. (6): 375-7.
2. Yaglou C P, Minard D. 1957. Control of heat casualties at military training centers. Amaarchives of Industrial Health. 16(4): 302-16.
3. Chinese standard. 2008. Classified standard of working in the hot environment (GB/T 4200-2008).
4. Chinese standard. 2010. Classification of occupational hazardsat ISO. 2004. Ergonomics of the thermal environment – Determination of metabolic rate (ISO 8996:2004).
6. ISO. 2017. Ergonomics of the thermal environment – Assessment of heat stress using the WBGT (wet blub globe temperature) index (ISO 7243:2017).
7. Montoye H J, Taylor H L. 1984. Measurement of physical activity in population studies: a review. Human Biology. 56(2): 195-216.
8. Li Y Q. 2016. Dynamic human heat strain prediction in hot working environment (medium metabolic rate and above). Ph.D. Thesis, Chongqing University, China.
9. Xu M N. 2013. Research on the adaptation of prediction heat strain model and safe exposure time evaluation. Ph.D. Thesis, Chongqing University, China.
10. Yaglou C P, Minard D. 1957. Control of heat casualties at military training centers. Amaarchives of Industrial Health. 16(4): 302-16.
11. Yao, R., B. Li, and J. Liu, *A theoretical adaptive model of thermal comfort - Adaptive Predicted Mean Vote (aPMV).* Building and Environment, 2009. **44**(10): p. 2089-2096.
12. Yang, Y., et al., *A study of adaptive thermal comfort in a well-controlled climate chamber.* Applied Thermal Engineering, 2015. **76**: p. 283-291

Study of indoor PM2.5 distribution characteristics and optimal location for its monitoring in an office

Guoxiang Hong[1, 2], *Chunhui* Liao[1, 2], *Hong* Liu[1, 2]

[1]Joint International Research Laboratory of Green Buildings and Built Environments (Ministry of Education), Chongqing University, Chongqing 400045, China
[2] National Centre for International Research of Low-carbon and Green Buildings (Ministry of Science and Technology), Chongqing University, Chongqing 400045, China

Abstract. With the appearance of the word "haze" in China, PM2.5 (particulate matter with aerodynamic diameter less than 2.5 μm) that can enter occupants' lungs has become a public topic of discussion. Today, the indoor PM2.5 or fine particulate concentration has become one of important factors affecting indoor air quality (IAQ). How to properly monitor indoor PM2.5 is an urgent issue to be discussed and solved. At present, sampling is adopted to know PM2.5 concentration in a room, and Chinese related standard required the sampling time for indoor PM2.5 is at least 8 hours. However, the sampling method takes too much time, and the HVAC system cannot react in real time such as increasing the fresh air volume with increase of indoor PM2.5 concentration. So, there is a great need to find an optimal location for continuous PM2.5 monitoring. Before finding the monitoring point, the spatial and temporal distribution characteristics of indoor PM2.5 concentration are needed to be known. Computational fluid dynamics (CFD) can be used to simulate airflow and dispersion of PM2.5 in rooms with different scales, functions and ventilations. This paper will contribute to find the optimal location which could preferably describe indoor PM2.5 concentration in an office combined with experimental research and CFD simulation. In short, the aim of the paper is to reveal the spatial-temporal characteristics of indoor PM2.5 concentration distribution and optimize the layout of PM2.5 monitoring points for air conditioning systems to better control indoor contaminate PM2.5 concentration.

1 Introduction

The first time that the U.S. environmental protection agency (USEPA) [1] established PM2.5 was in 1997 to protect public health. Due to the rapid industrialization and high energy consumption, PM2.5 pollution is gradually exploding. It can increase the incidence of various respiratory diseases and occupant mortality rates, as well as increase of the energy consumption in heating, ventilation, and air conditioning (HVAC) systems. However, nowadays people have spent as much as 80% to 90% hours indoors, such homes, offices, and transport vehicles [2-4]. Meanwhile, indoor air quality (IAQ), which have a directly significant influence on health as well as efficiency of study and work [5-7], is becoming worse and worse. To solve the problem produced by indoor PM2.5, it is essential to study the indoor PM2.5 concentration distribution characteristics.

Since PM2.5 attached public attention, many researchers studied on indoor PM2.5. People, who have jobs, spend most time of their day in the office, and there are many sources that produce particulate matter such as outdoor sources, human activities, printer and other electrical appliances. This study combines experimental research and CFD simulation to the spatial and temporal distribution characteristics of indoor PM2.5 concentration in an office. Some previous works have done some studies on particle emissions and its concentration of printers in the office room [8,9]. And other works used the CFD to simulate indoor PM2.5 distribution in different ventilated ways [10-16]. But, there are few studies to find the optimal location for monitoring, so, this is where this study is deserved.

This study focused on the indoor PM2.5 distribution characteristics and optimal location for its monitoring in an office. The indoor PM2.5 concentration deserved to study deeply, not only to find the optimal monitoring location, and for the occupants' and workers' health. Although the analysis about indoor PM2.5 distribution and the selection of monitoring location in this paper may be superficial, it is meaningful to start studies in this field of indoor PM2.5.

2 Experimental method

2.1 Monitoring site

A typical office in Chongqing University was chosen as the monitoring site.

2.2 Instrumentation

PM2.5 concentration was measured by laser sensor produced by Panteng Company in China, table 1 shows its performance parameter.

Table 1. characteristics of PM2.5 sensor

Parameter	Index	Unit
Particle measurement range	0.3-1.0; 1.0-2.5; 2.5-10	μm
Particle count efficiency	50%@0.3μm 98%@≥0.5μm	
Particle mass concentration effective range (PM2.5 standard value)	0-500	μg/m^3
Particle mass concentration resolution	1	μg/m^3
Particle mass concentration consistency (PM2.5 standard value)	±10%@100-500μg/m^3 ±10μg/m^3@0-100μg/m^3	
Weighing volume	0.1	L
Single response time	<1	s
Integrated response time	≤10	s

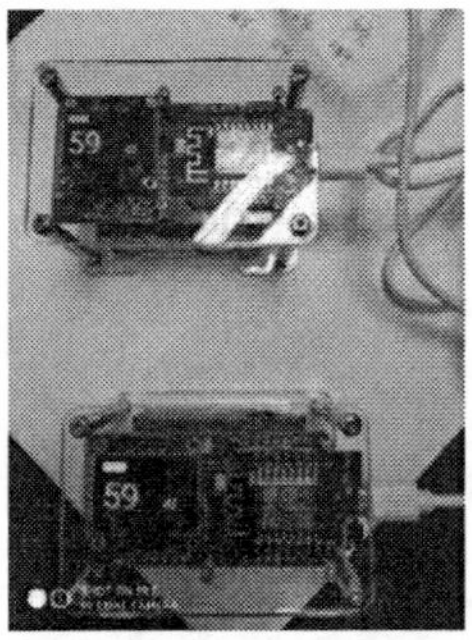

Figure 1. Laser sensor for PM2.5 monitoring

Temperature data acquisition was obtained with the Agilent 34980A Multifunction Intelligent Data Collector with a T-type thermocouple. In order to ensure the accuracy of the data, the thermocouple is calibrated using the constant temperature water tank method.

The wind speed data of the supply and return air is collected by the Swema 3000 anemometer. The measurement method of wind speed refers to the "Code for Construction Technology and Quality Acceptance of Ventilation and Air Conditioning Engineering" GB 50019-2003, which adopts the fixed-point measurement method.

Source of PM2.5 in this study is incense. The burnout time of the incense is about three hours, and the emission rate of PM2.5 is 5.33μg/m^3.

2.3 Arrangement of PM2.5 monitoring point

In this experiment, there are ten locations for PM2.5 monitoring totally. Five locations are in the space and other five points are near the ceiling or the west wall. The five points in the space are evenly distributed across the diagonal and their height is 1.2 m in the breathing zone. The other five points are 0.1 m away from the wall or ceiling and the specific arrangement for these points is shown in Figure 2.

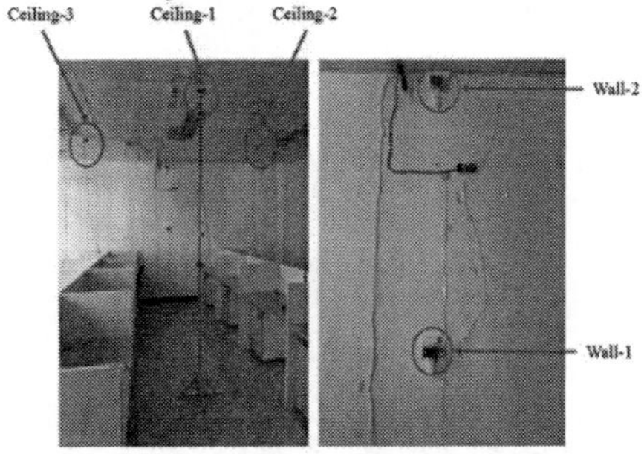

Figure 2. Arrangement for 5 points near the ceiling or west wall

2.4 Data processing and analysis

All statistical analyses (correlation, regression, t-test) were conducted using Excel (Microsoft). When the distribution of the data was not a normal distribution, the robust analysis (trim off the maximum and minimum) was employed.

3 CFD modeling

3.1 Model for an office established

According to the structure parameters of the room in actual engineering, the simulation model on the distribution of PM2.5 in indoor air is established in the condition of air conditioning. Figure 3 shows the office model in the air conditioning ventilation and specifications of the office are listed in Table 2. The model is based on the actual room size, meanwhile some indoor display was simplified.

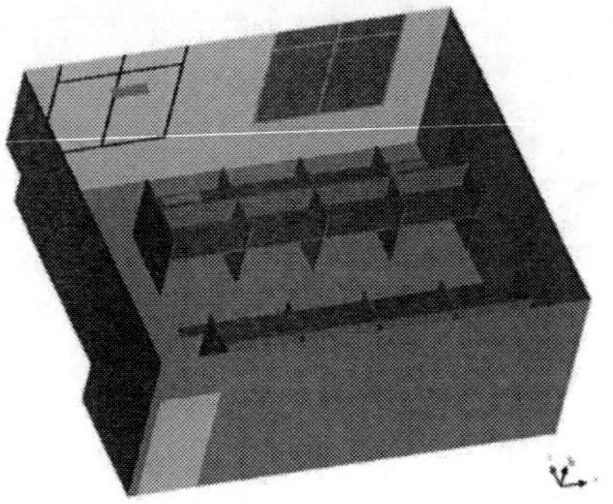

Figure 3. A schematic view the office used for full-scale experiments and simulations

Table 2. Specifications of the office

Ventilation	Dimensions(m)			velocity of supply vent	Particle size
	Length	Width	Height	(m/s)	μm
Ceiling supply (fan coil)	7	6	2.9	5	≤2.5

3.2 Simulation method

In this study, the ANSYS 15.0 was adopted to analyze the indoor PM2.5 distribution characteristics. Considering forces on the PM2.5, the Basset history, the pressure gradient and the virtual mass are ignored because they are infinitesimal and negligible compared to the drag force [12]. Meanwhile the PM2.5 is assumed to be the spherical particulate matters with a diameter less than 2.5 μm [14]. This paper adopts the Euler-Lagrange model to track the movements of PM2.5. The continuous phase uses the k-ε turbulence model to couple calculation with discrete phase. However, the standard k-ε turbulence model and the Realizable k-ε turbulence model were both adopted in different studies of indoor PM2.5. So, this paper will make comparison of two k-ε turbulence models and obtain the optimal model for the office. In this simulation, the coupling calculation with discrete will be occurred when the continuous phase is calculated every two iteration step. For the analyses about the indoor PM2.5 concentration distribution characteristics, the inlet of the office adopts the velocity boundary for the continuous phase. At the same time, the escape boundary for the discrete phase is adopted in that. The outlet adopts the outflow boundary for the continuous phase and the escape boundary for the discrete phase. In addition, the wall, floor, ceiling adopt trap boundary for the discrete phase.

3.3 Result of simulations and discussion

Figure 4 depicts the mass concentration distribution of indoor PM2.5 at different section of X=5.2 m, Z=1.2 m, Z=ceiling under two different turbulence models. The results show that the indoor PM2.5 diffuses uniformly in the standard k-ε turbulence model, especially in the Z=ceiling. So, the standard k-ε turbulence model could describe indoor PM2.5 dispersion better for the office model. Figure 5 shows three counters of PM2.5 mass concentration under the standard k-ε turbulence model. In Figure 5, the mass concentration of PM2.5 in X=3.5 m is larger than the mass concentration in X=5.2 m located around inlet, so the indoor PM2.5 concentration is directly influenced by velocity. At the same time, these results indicate that the optimal location for PM2.5 monitoring could be found to represent the mean value of indoor PM2.5 concentration in the breathing zone.

standard k-ε turbulence model

X=5.2 m

Realizable k-ε turbulence model

X=5.2 m

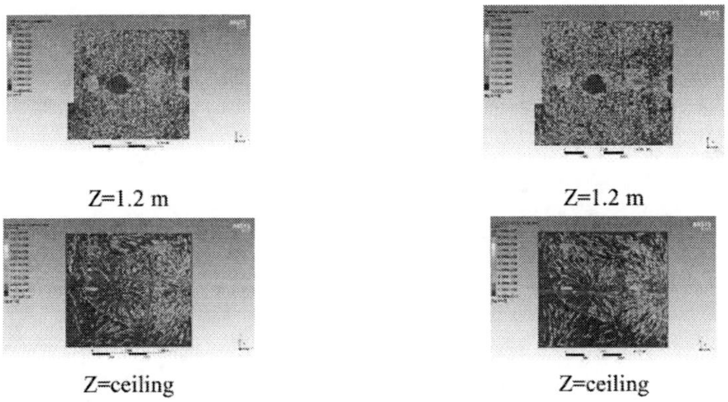

Figure 4. Comparison of simulation results about indoor PM2.5 concentration distribution in three vertical views

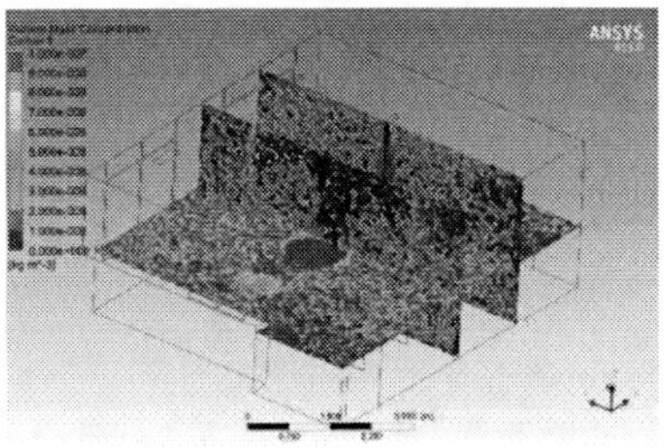

Figure 5. Counters of PM2.5 concentration in standard k-ε turbulence model

4 Experimental results and discussion

4.1 Experimental conditions and results

Experiment in the office has eight conditions. The six conditions monitor the background mass concentration of PM2.5 with three different velocity (2.0 m/s, 3.5 m/s, 5.0 m/s) and two different temperature (22 and 28 °C) in inlet. One condition combines the fan coil and fresh air with the velocity=5.0 m/s and temperature=22 °C in inlet. Another condition burns the incense for the PM2.5 source with the velocity=5.0 m/s and temperature=22 °C in inlet and no fresh air. Figure 6 shows curves of mean PM2.5 concentration of five points in space under 6 conditions with 60 minutes when the air conditioning is stable. Figure 7 depicts that the mean PM2.5 concentration in space and other PM2.5 concentration near the ceiling and wall changes with time after the air conditioning is opened in two hours.

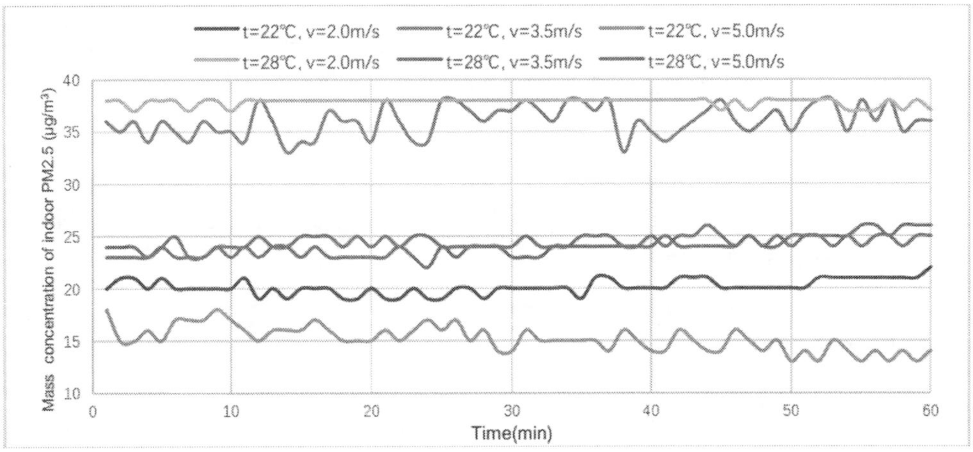

Figure 6. Mean value of background PM2.5 concentration in 6 conditions (no sources and fresh air supply)

https://doi.org/10.1051/e3sconf/20198003006

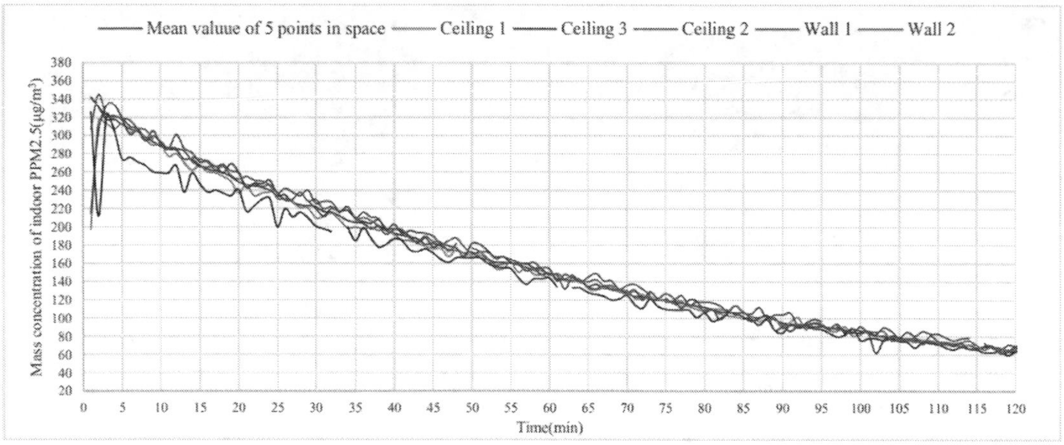

Figure 7. Comparison of mean PM2.5 concentration in space and other 5 point's PM2.5 concentration near ceiling and wall

4.2 Discussion

Figure 6 shows that mean mass concentration of indoor PM2.5 is steady if there is no evident source which can generate PM2.5. The indoor PM2.5 concentration has a great correlation with the outdoor PM2.5 concentration, which can be known by Figure 6, where the outdoor PM2.5 concentration is larger in "t=22 °C, v=3.5 m/s" and "t=28 °C, v=2.0 m/s" than in other conditions. Figure 7 shows that the mean value of indoor PM2.5 in space is similar with the indoor PM2.5 concentration near ceiling or wall. So, it indicates that the 5 locations near ceiling and wall can represent the mean value in space in some extent again.

This study makes the comparison between indoor PM2.5 concentration of 5 points near the ceiling and wall with the mean value of PM2.5 concentration in space. Table 3 shows the deviation of location near ceiling or wall from the mean value. Ceiling 3, ceiling 3, ceiling 1 perform better in the velocity=2.0 m/s, 3.5 m/s, 5.0 m/s respectively.

Table 3. Deviation of location near ceiling or wall from the mean value

Deviation from the mean value [mean value of deviation (minimum value of deviation, maximum value of deviation)]					
	Velocity=2.0m/s				
Temperature	Ceiling 1	Ceiling 2	Ceiling 3	Wall 1	Wall 2
22°C	-0.27% (-9.52%, 10.53%)	-2.86% (-10.00%, 10.53%)	-1.25% (-14.29%,15.79%)	6.45% (-4.76%, 26.32%)	0.58% (-14.29%, 10.53%)
28°C	-5.76% (-10.53%, 0.00%)	-6.47% (-10.53%, -2.63%)	-3.96% (-10.53%,0.00%)	-3.21% (-7.89%, 2.70%)	-11.4% (-15.79%, -5.41%)
	Velocity=3.5m/s				
Temperature	Ceiling 1	Ceiling 2	Ceiling 3	Wall 1	Wall 2
22°C	-6.75% (-11.43%, -2.86%)	-3.66% (-11.43%, 2.86%)	-0.35% (-5.71%, 5.88%)	4.84% (-2.86%, 9.09%)	-7.05% (-11.43%, -2.86%)
28°C	-0.13% (-8.70%, 13.64%)	-4.24% (-12.50%, 8.70%)	-2.16% (-12.50%, 4.55%)	-1.22% (-12.50%, 8.70%)	-6.46% (-12.50%, 0.00%)
	Velocity=5.0m/s				
Temperature	Ceiling 1	Ceiling 2	Ceiling 3	Wall 1	Wall 2
22°C	0.16% (-14.29%,14.29%)	-3.51% (-14.29%, 7.69%)	-2.37% (-14.29%, 14.29%)	3.95% (-7.14%, 20.00%)	-3.88% (-15.38%, 13.33%)
28°C	-2.68% (-16.00%, 12.50%)	-2.73% (-12.00%, 8.33%)	-1.80% (-16.00%, 8.33%)	-0.83% (-12.00%, 8.33%)	-7.20% (-16.00%,0.00%)

Condition 7: combing fan coil and fresh air with the velocity=5.0 m/s and temperature=22 °C in inlet. Figure 8 shows that mass concentration of PM2.5 near ceiling or wall and mean PM2.5 mass concentration of 5 points in space have a good correlation. The point "Ceiling 1" performs the best correlation among 5 points, which can be seen from Table 4 and its value of R^2 is 0.8854.

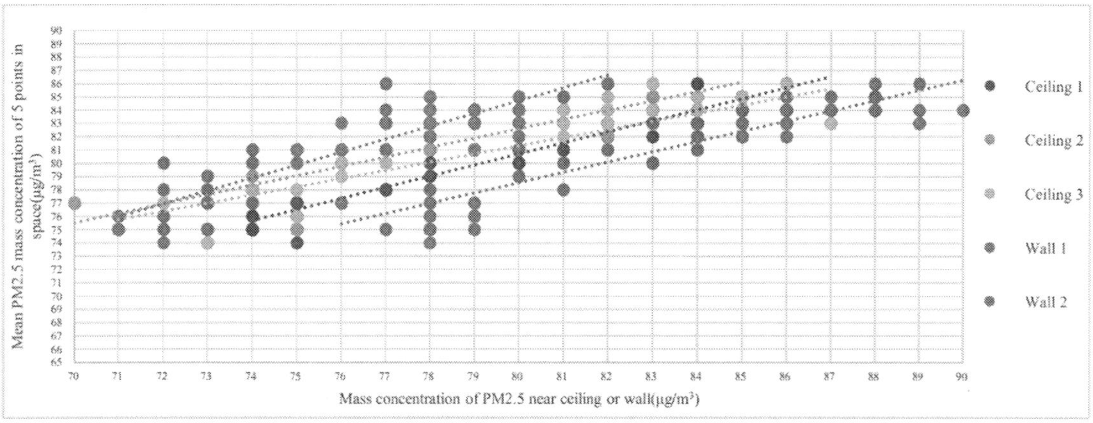

Figure 8. Linear regression between mass concentration of PM2.5 near ceiling or wall and mean value in space under condition 7

Table 4. Results of linear regression on condition 7

	Slope	Intercept	R^2 value
Ceiling 1	0.8332	14.04	0.8854
Ceiling 2	0.7049	26.197	0.7928
Ceiling 3	0.6149	32.115	0.7942
Wall 1	0.7732	16.684	0.8088
Wall 2	0.9687	7.2234	0.8047

Condition 8: burning the incense for the PM2.5 source with the velocity=5.0 m/s and temperature=22 °C in inlet and no fresh air. In this condition, the linear relationship between mean PM2.5 mass concentration in space and PM2.5 mass concentration of five points near wall or ceiling performs perfectly and the point "Wall 1" performs the best correlation among 5 points, which can be seen from Table 5 and its value of R^2 is 0.9959.

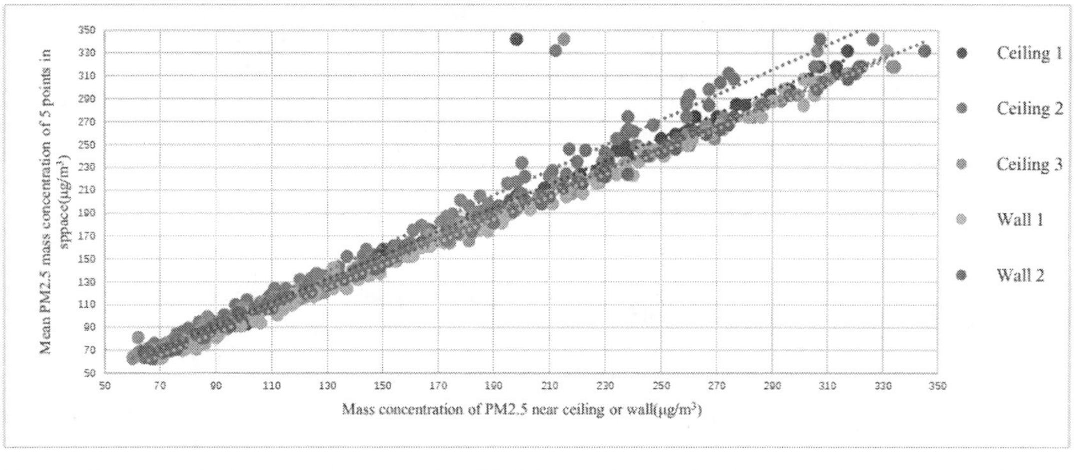

Figure 9. Linear regression between mass concentration of PM2.5 near ceiling or wall and mean value in space under condition 8

Table 5. results of linear regression on condition 8

	Slope	Intercept	R^2 value
Ceiling 1	1.0312	-1.6316	0.9685
Ceiling 2	0.986	1.8973	0.9705
Ceiling 3	1.0999	-3.6046	0.977
Wall 1	0.9924	-4.3395	0.9959
Wall 2	0.9832	0.8746	0.9958

5 Conclusion

After analysis, the following conclusions in this paper can be drawn:

1. When the indoor PM2.5 concentration is attenuated, the average value of the five monitoring points in the

spatial respiration zone has a large linear correlation with the value of the wall monitoring point. From this paper, the point "Ceiling 1" and "Wall 1" would be the optimal locations for the monitoring of PM2.5 concentration in this office.

2. When the indoor environment parameters in the air conditioning are stable, the average value of the five monitoring points in the space breathing zone is close to the value of some wall monitoring points.

At present, some domestic scholars and foreign scholars have devoted themselves to the study of indoor PM2.5 distribution characteristics under different air-conditioning systems. However, few researches have proposed reasonable monitoring points or monitoring criteria for indoor PM2.5. Therefore, this study is trying to fill the gap in this field.

By analyzing the indoor PM2.5 concentration distribution characteristics in the air conditioning system and reading the relevant references, the following innovations worth studying can be made:

(1) Conduct a typical personnel survey to analyze the effects of typical indoor personnel behavior on PM2.5 monitoring results and propose an indoor PM2.5 concentration distribution prediction model combined with experimental research and CFD simulation.

(2) Propose the air conditioning scheme optimization and design method based on indoor personnel PM2.5 inhalation risk as an evaluation index.

Acknowledgements

This research is supported by the National Key Research and Development Program of China (Grant No. 2017YFC0702704), the Fundamental Research Funds for the Central Universities (Project No. 106112015CDJXY210007), and the 111 Project (No.B13041)

References

1. Federal-Register. National ambient air quality standards for particulate matter. Fed. Regist, 1997, **62** (138).
2. Klepeis, N.E., Nelson, W.C., Ott, W.R., Robinson, J.P., Tsang, A.M., Switzer, P., et al.. The national human activity pattern survey (nhaps): a resource for assessing exposure to environmental pollutants[J]. Expo. Anal. Environ. Epidemiol, 2001 **11** (3): 231-252.
3. Leech, J.A., Nelson, W.C., Burnett, R.T., Aaron, S., Raizenne, M.E.. It's about time: a comparison of Canadian and American time-activity patterns[J]. Expo. Anal. Environ. Epidemiol, 2002 **12**: 427-432.
4. Robinson, J., Nelson, W.C.. National Human Activity Pattern Survey Data Base. USEPA, Research Triangle Park, NC, 1995.
5. Deng, Q., Lu, C., Ou, C., Chen, L., Yuan, H.. Preconceptional, prenatal and postnatal exposure to outdoor and indoor environmental factors on allergic diseases/symptoms in preschool children[J]. Chemosphere, 2016, **152**: 459-467.
6. Lu, C., Deng, Q., Li, Y., Sundell, J., Norbäck, D.. Outdoor air pollution, meteorological conditions and indoor factors in dwellings in relation to sick building syndrome (sbs) among adults in China[Sci]. Total Environ, 2016, **560-561**: 186-196.
7. Norbäck, D., Hashim, J.H., Markowicz, P., Cai, G., Hashim, Z., Ali, F., et al.. Endotoxin, ergosterol, muramic acid and fungal dna in dust from schools in Johor Bahru, Malaysia—associations with rhinitis and sick building syndrome (sbs) in junior high school students [Sci]. Total Environ, 2016, **545-546**: 95-103.
8. Tunga Salthammer, Tobias Schripp, Erik Uhde, Michael Wensing. Aerosols generated by hardcopy devices and other electrical appliance[J]. Environment Pollution, 2012, **169**: 167-174.
9. Antti J. Koivistoa, Tareq Hussein, Raimo Niemelä, Timo Tuomi, Kaarle Hämeri. Impact of particle emissions of new laser printers on modeled office room[J]. Atmospheric Environment, 2010, **44**: 2140-2146.
10. K C Chung. Three-dimensional analysis of airflow and contaminant particle transport in a partitioned enclosure[J]. Building and environment, 1999, **34**: 7-17.
11. Z. Zhang, Q. Chen. Experimental measurements and numerical simulations of particle transport and distribution in ventilated rooms[J]. Atmospheric environment, 2006, **40**: 3396-3408.
12. Bin Zhao, Ying Zhang, Xianting Li, Xudong Yang, Dongtao Huang. Comparison of the modelling and the experimental results on[J]. Building and environment, 2004, **39**: 1-8.
13. Z.F. Tian, J.Y. Tu, G.H. Yeoh, R.K.K. Yuen. On the numerical study of contaminant particle concentration in indoor airflow[J]. Building and environment, 2006, **41**: 1504-1514.
14. Yuting Ma, Yiqiang Jiang, Lan Li. Numerical Simulation of PM2.5 Distribution in Indoor Air[J]. Procedia engineering, 2015, **121**: 1939-1947.
15. Zong Qingsong. Studies on Distribution and Deposition of Indoor Particulate Matter [D]. Anhui University of Technology, 2012.
16. Chen Jiwei. Numerical Simulation of Airflow and Tracks of Particles in Air-conditioned Room[D]. Nanjing University of Science & Technology, 2007.

Acceptable surface temperature of floor radiant heating system based on thermal comfort study in southern China

Xingyu Lu [1, 2], *Hong* Liu [1, 2], and *Yuxin* Wu [1, 2]

[1] Joint International Research Laboratory of Green Buildings and Built Environments (Ministry of Education), Chongqing University, Chongqing, China
[2] National Centre for International Research of Low-carbon and Green Buildings (Ministry of Science and Technology), Chongqing University, Chongqing, China

Abstract. In winter, people's demand for heating is stronger and stronger in southern China, and the floor radiant heating system is more and more popular. However, there are no suitable guidelines or standards for the floor temperature of the heating system in this area. The insulation performance of buildings in southern is not as good as other area which have central heating system. So the acceptable floor temperature suitable for this area needed to be studied.12 healthy college students participated as samples in this experiment. The floor surface temperature was controlled by varying the temperature of water flowing underneath the floor. The main conclusions were as follows: 1) the floor surface temperature directly affected the skin temperature of the foot and the thermal comfort of the foot. There was a significant statistical relationship between the floor surface temperature and the overall thermal sensation. 2) The acceptable floor temperature ranged from 26.1 °C to 34.3 °C for sitting positions and 24.6 °C ~34.7 °C for standing positions. 3) Considering the head thermal comfort and the health effects of the cumulative effect of long-term heat exposure, the recommended upper limit of the floor temperature in this experiment is 31°C.

1 Introduction

In southern China, with the frequent occurrence of extreme cold weather, people's demand for heating is getting stronger and stronger. However, the insulation performance of building walls is not as good as that of buildings in the northern regions where central heating is used for a long time, and the ratio of windows to walls is bigger. So the acceptable floor temperature of occupants based on thermal comfort may be not the same. Low temperature hot water floor radiant heating system which appeared in the early 20th century [1] has received more and more attention especially by the southern people due to the low temperature of hot water, good comfort and energy saving. However, how to create a comfortable indoor environment by the use of floor radiant heating and formulate industry standards remain to be solved.

The European, American, and international standards [2-5]governing the temperature range of radiant floor heating are mostly in the range of 19 to 29 °C. The sources of data are mainly those of Olesen et al [6-7]. from the Technical University of Denmark in the 1970s-1980s. While the Chinese standard is 24/25~26/27°C, the upper limit is 28/29 °C[8]. The data source can be considered as part of the reference to foreign early research data. However, due to differences in races, living environment and living habits, Asian populations are more tolerant of heat, so the acceptable comfortable floor temperature should be higher than that of European and American people. The Japanese and Korean scholars have confirmed that. In Korea, there is a traditional floor heating method called "ONDOL", which has a comfortable floor surface temperature range of 33.6~38.8 °C[9]. The upper and lower temperature limits when sitting are related to the material of the floor. The neutral temperature of the human body is 32.6 °C[10]. When standing with barefoot, the floor temperature range is 27~33 °C[11]. Japanese scholars have concluded that the upper and lower limits of acceptable temperature are 31 ° C and 25 ° C, respectively, and the optimal floor temperature range is 26 ~ 30 ° C[12]. It can be seen that the temperature range obtained by research in Asian regions such as Korea and Japan is higher than the international standard (19~29 °C).

When using the floor radiant heating, floor contacts the foot directly, and the comfortable air temperature required for different part of human body is different [13]. In the hot environment, head is sensitive to heat discomfort, while feet and hands can easily feel cold and uncomfortable in the cold environment[14]. If staying on the warmer floor for a long time, when the skin temperature of feet is greater than 34 °C, the blood flow is significantly increased [15]. In addition, the increase of skin temperature will relax the capillaries, promote blood circulation, and enhance the oxygen supply of various tissues and organs of the human body [16-17]. Nevertheless, if the ambient temperature around the feet is at a high level for a long time, it is not only harmful to patients with cardiovascular disease and diabetes [18], but also has adverse effects on healthy people [19]. Therefore, in determining the upper limit of the floor temperature, both the thermal comfort requirement and the long-term cumulative effect on health should be considered.

Floor radiant heating creates an unsteady indoor environment, comfort temperature is not equal to neutral temperature. That is to say, people who undergo a number of cold environments may prefer warmer floor temperature instead of neutral floor temperature[20-21]. Therefore, as far as the southern region is concerned, it is possible to further increase the upper limit of 29 °C in the current Chinese standard. Thus, the study focused on the actual heating environment in southern buildings. From the perspective of thermal comfort, combined with the frequency of sick symptoms, an acceptable floor temperature range adapted to the southern region was proposed.

2 Research methods

2.1. Experimental platform

The experiment was carried out in a capillary radiation laboratory with a floor area of 3 m*1.5 m and a height of 3 m. The air source heat pump unit and the heating water tank were used to prepare hot water, and the temperature probe was arranged in the water tank to monitor the water temperature in time. The hot water entered the capillary network under the floor in the room for radiant heating, as Figure 1 shows. Floor material was polystyrene board, shown in Figure 2. a), and the wall was made of ordinary brick wall with thermal insulation material, which was similar to the common structure of architecture in southern China.

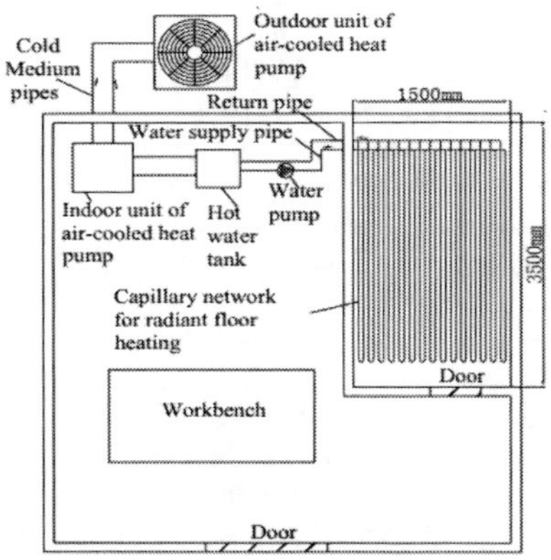

Fig.1. Diagram of the experimental platform.

The indoor air vertical temperature and floor surface temperature were measured by an Agilent 34970A data acquisition instrument with copper-constantan thermocouples. The measurement accuracy after water bath calibration was ±0.1 °C. The measuring points on the floor surface are arranged in three equal parts of the diagonal[8], a total of five (a,b,c,d,e) as Figure 2. b) shows. Point A, B and C were the position of subjects. The height of the vertical temperature measurement points included 0.1 m (ankle), 0.6 m (knee), 1.1 m (head by sitting), 1.7 m (head by standing), and 2.5 m (top of the room), as Figure 2. c) shows. The skin temperature of human body and indoor air relative humidity were measured by HOBO UX100-011 temperature and humidity self-recording instrument, and the measurement accuracy was ±0.21 °C; ±2.5%. The measurement points of skin temperature included upper arm, chest, anterior thigh, interior calf, instep and ankle. The measurement time of all parameters were 90 minutes.

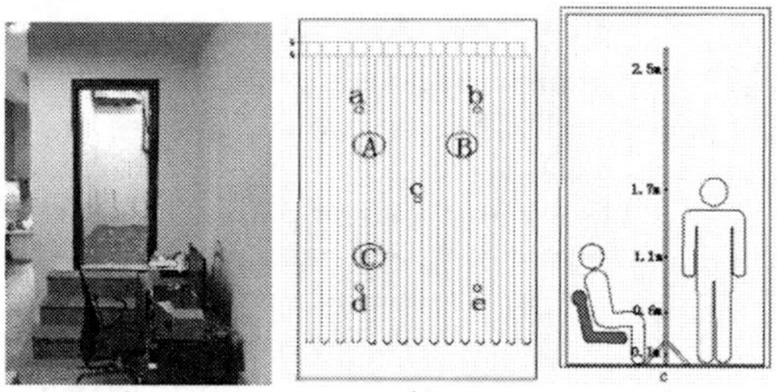

a) b) c)

Fig.2. Diagram of Experimental details: a) exterior of the Capillary radiation room, b) measuring points on the floor, c) measuring points of vertical temperature

2.2 Experimental conditions and procedure

The experiment was carried out in January 2018 in Chongqing, China. Chongqing is a typical city in southern China, which is located in the hot summer and cold winter climate zone. The outdoor temperature during the experiment was from 2 °C to 8 °C. Experimental conditions were designed with six different floor temperatures. The floor surface temperature was controlled by varying the temperature of water (t_w) flowing underneath the floor, and it ranged from t_w= 30 °C to 55 °C at 5 °C intervals. The background temperature is maintained at a neutral temperature of 20 to 22 °C and relative humidity of 50%.

Before the experiment, the subjects were asked to be familiar with the experiment requirements and questionnaire in the natural ventilation room, clothing is about 1.2 clo.Then entering the floor heating room after eliminating the previous thermal experience by resting for about 20 minutes. At the beginning of the experiment, the subject took off his shoes and stepped on the floor wearing socks. Subjects could read books or use mobile phones, but are not allowed to communicate and discuss about thermal comfort. During the experiment, subjects were asked to fill out the questionnaire and measured physiological parameters at intervals of 5 min or 10 min.At the same time, the operator recorded environmental parameters, as shown in Figure 3.

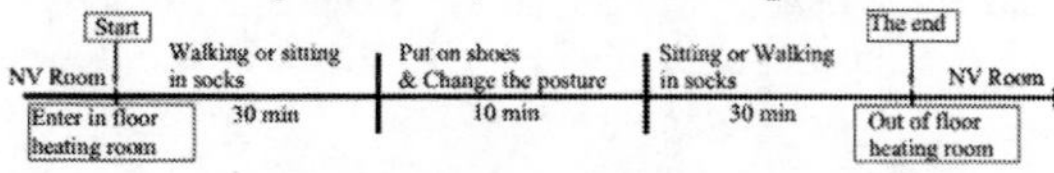

Fig.3. Experimental procedures

To avoid the effects of posture on thermal comfort, subjects were randomly assigned to two groups: first standing, sitting back, or sitting first and then standing. Due to the temperature drop of the hot water pipe network and the heat dissipation of floor material, the final floor surface temperature was averaged by the actual temperature measured by five points. Thus the actual floor surface temperature were used to analyse in the paper.

2.3 Subjects

Twelve healthy college students participated in the experiment, with a male to female ratio of 1:1. All of the subjects are currently living in Chongqing, China, most of them are born in southern China, only a few are from the central region, and more importantly, they have no thermal experience of central heating in winter. Their information is listed in Table 1. It can be seen that women will choose more diversified heating methods in winter, for instance, combining air conditioning and auxiliary heating equipment to achieve better thermal comfort.

Table 1. Information of Subjects

Group	Age	Height (cm)	Weight (kg)	Body fat rate (%)	Heating method in winter (Proportion %)
Female	23±1	161.7±5.2	52.9±6.0	26.2±3.9	Air conditioning heating (66.7%)
					Local heating (16.6%)
					Air conditioning heating + Local heating (16.6%)
Male	23±1	177.3±4.6	68.7±8.4	19.6±3.9	Air conditioning heating (66.7%)
					Local heating (33.3%)
All	23±1	169.5±9.2	60.8±10.7	22.9±5.1	Air conditioning heating (66.7%)
					Local heating (25%)
					Air conditioning heating + Local heating (8.3%)

2.4 Questionnaires

Questionnaires overed the thermal sensation vote (TSV), thermal pleasure vote (TPV) and a variety of sick symptom. Both the overall perception and local perception were evaluated, especially the heat perception of the feet.Thermal sensation was assessed using the ASHRAE seven-point thermal sensation scale -3 cold, -2 cool, -1 slightly cool, 0 neutral, +1 slightly warm, +2 warm, +3 hot. TPV: -3 very unpleasant, -2 unpleasant, -1 slightly unpleasant, 0 indifferent, +1 slightly pleasant, +2 pleasant, +3 very pleasant [22].Sick symptom including: headache, dizzy, foot pain, pedal numbness, fatigue and lethargy, which is evaluated by four level: No,slightly,obviously, strongly. In addition. The subjects were also asked whether it was difficult to concentrate and do not want to study.

3 Result and discussion

3.1 Thermal environment monitored

For the sake of analyzing whether the vertical temperature difference of the environment would cause thermal discomfort, the test was repeated four times for each working condition, and the surface temperature of the floor corresponding to the six different water supply temperatures was shown in the Table 2.

Table 2. Actual floor temperature

Water supply temperature(°C)	Floor surface temperature(°C)
55	36.4±0.3
50	33.9±0.5
45	31.8±0.3
40	29.6±0.4
35	27.4±0.4
30	25.8±0.2

The temperature at different vertical heights is shown in the figure 4.It can be seen that when tw=55°C, tw=50°C, tw=45°C, the air in the room was hot below but cold above. While when tw=40°C, tw=35°C, and tw=30°C, the room was cold below but hot above. However, the vertical temperature difference (temperature difference between the head and the foot) of all conditions didn't exceed 3 °C, which met the thermal comfort requirement [3].

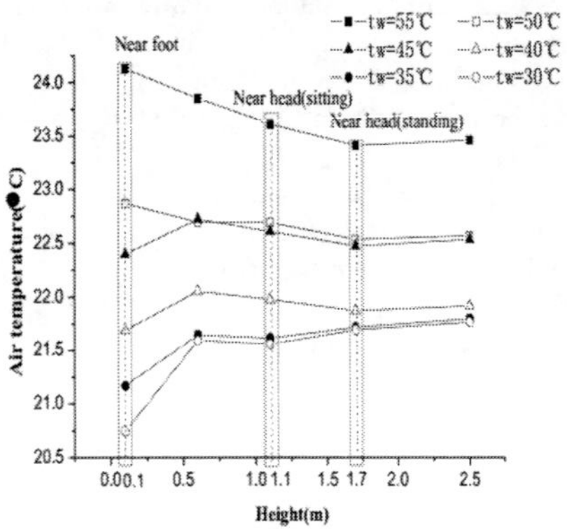

Fig.4. Vertical temperature distribution of indoor air

3.2 Skin temperature, TSV and TPV of foot

Took the average of the instep and ankle temperature as the skin temperature of the foot. The skin temperature of foot when sitting was higher than standing, and it showed a more significant correlation with the floor temperature. When the floor temperature was less than 34.5 ° C, the TSV of standing was hotter than sitting. While after 34.5° C, the thermal sensation of sitting was hotter than that of standing, and the rising speed of TSV increased with the floor temperature rising. The TPV of standing was better than that of sitting before the floor temperature was 34℃,but it obviously decreased after 34℃, and the TPV of sitting increased as the floor temperature increased.

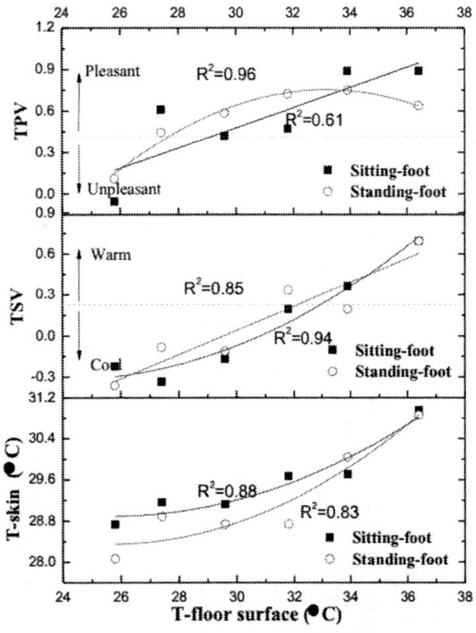

Fig.5. Skin temperature, TSV and TPV of foot with floor temperature

3.3 MST, TSV and TPV of overall and head

The mean skin temperature(MST) was calculated by weighting the local skin temperature. The formula[23] is as follows :

$$MST=0.3T_{sk,ua} + 0.3T_{sk,ch} + 0.2T_{sk,at} +0.2\ T_{sk,ac} \qquad (1)$$

Where $T_{sk,ua}$ is temperature of upper arm, °C. $T_{sk,ch}$ is temperature of chest, °C. $T_{sk,at}$ is temperature of anterior thigh, °C. $T_{sk,ac}$ is temperature of anterior calf, °C.

Head is the farthest part from the heating floor. Furthermore, the TSV and TPV of head have a greater impact on the work and learning efficiency of occupiers. As can be seen from Figure 6, the MST changed little, and the difference was within 1.5 °C. There was a significant positive correlation between the TSV and the floor temperature. When sitting, the TSV of overall and head both rose slowly with the increased of floor temperature, and the rate of change was basically unchanged. However, in standing condition, the increase rate of overall TSV was faster first and then slower, while the head TSV showed a slowly rising trend. In both postures, overall TPV and head TPV were relatively less correlated with floor temperature, especially overall. For the head, the TPV decreased as the floor temperature increased. When the floor temperature exceeds 31 °C, the rate of decline in standing position was significantly accelerated.

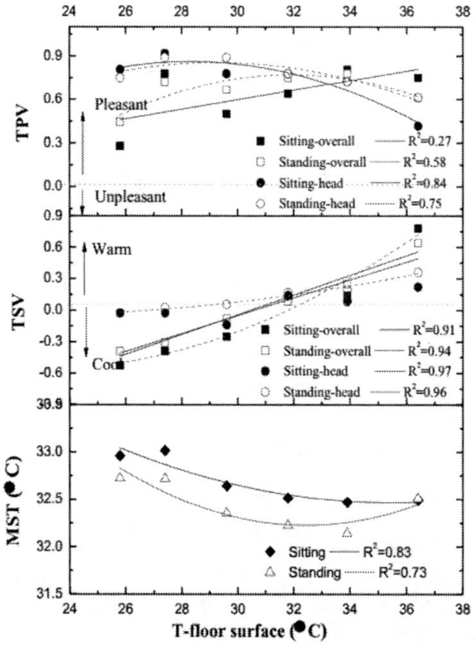

Fig.6. MST, TSV and TPV of overall and head with floor temperature

 https://doi.org/10.1051/e3sconf/20198003007

3.4 Percentage Dissatisfied (PD) of cool and warm

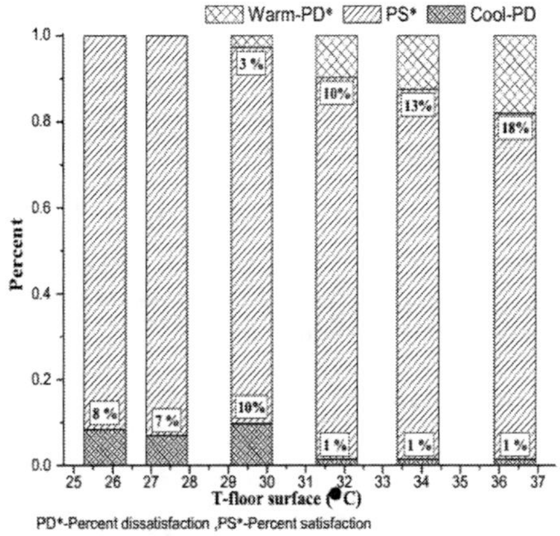

Fig.7. PD of cool and warm with floor temperature by sitting

Thermal discomfort is first shown by people who have a thermal sensation higher than +2 or below -2[24]. Therefore, it was determined to be dissatisfied when felt obvious uncomfortable in this experiment. The percentage dissatisfied of warm(warm-PD) was defined as the percentage of subject's TSV was greater than or equal to +2. Analogous to the warm-PD, cool-PD meant subject's TSV was less than or equal to -2. In both postures, the rate of cold dissatisfaction decreased as the temperature increased. When the floor temperature exceeded 29.5 °C, the thermal dissatisfaction rate appeared and rose. Moreover, the cold dissatisfaction rate of sitting posture was obviously higher than that of standing posture, and the thermal dissatisfaction rate is also slightly higher.

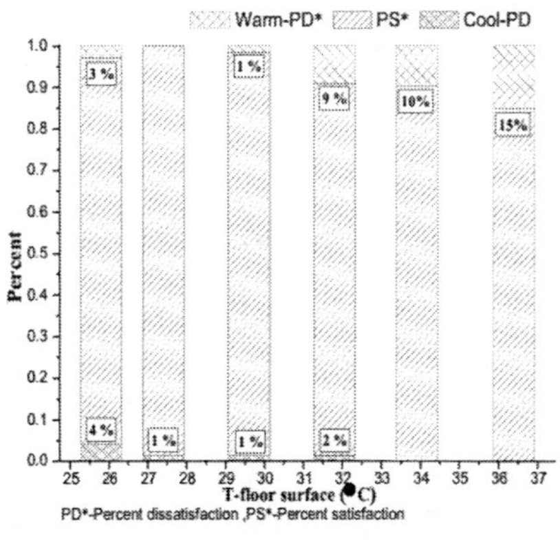

Fig.8. PD of cool and warm with floor temperature by standing

3.5 Sick symptom

In the experiment, headache, dizzy were unified as head discomfort. Foot pain, pedal numbness were foot discomfort. Fatigue and lethargy were poor mental state. The results are shown in the table 3.

Table 3.a) Sick symptom with floor temperature by sitting

Sick symptom	Level	Floor temperature(°C) 25.8	27.4	29.6	31.8	33.9	36.4
Head symptom	O*	0	0	0	0	0	0
	S*	0	0	0.01	0.01	0.07	0.1
	N*	1	1	0.99	0.99	0.93	0.9
Foot symptom	O*	0	0	0	0	0	0
	S*	0.07	0	0.07	0.04	0.03	0.03
	N*	0.93	1	0.93	0.96	0.97	0.97
Poor mental state	O*	0	0.01	0	0.03	0.04	0.03
	S*	0.24	0.29	0.33	0.31	0.32	0.1
	N*	0.76	0.69	0.67	0.67	0.64	0.88
Difficult to concentrate and do not want to study	O*	0	0.01	0	0	0.04	0.01
	S*	0.29	0.28	0.19	0.28	0.31	0.17
	N*	0.71	0.71	0.81	0.72	0.65	0.82

O*- Obviously, S* -Slightly, N*- No

Table 3.b) Sick symptom with floor temperature by standing

Sick symptom	Level	Floor temperature(°C) 25.8	27.4	29.6	31.8	33.9	36.4
Head symptom	O*	0	0	0	0	0	0
	S*	0	0	0.07	0	0.03	0.06
	N*	1	1	0.93	1	0.97	0.94
Foot symptom	O*	0.01	0	0	0.01	0.01	0.06
	S*	0.08	0.07	0.1	0.07	0.11	0.08
	N*	0.9	0.93	0.9	0.92	0.88	0.86
Poor mental state	O*	0	0	0	0.07	0	0.01
	S*	0.22	0.26	0.36	0.22	0.26	0.22
	N*	0.78	0.74	0.64	0.71	0.74	0.76
Difficult to concentrate and do not want to study	O*	0	0	0	0.04	0.01	0
	S*	0.21	0.26	0.29	0.22	0.4	0.18
	N*	0.79	0.74	0.71	0.74	0.58	0.82

O*- Obviously, S* -Slightly, N*- No

In the standing position, as the temperature decreases, the head discomfort was reduced. Simultaneously, the foot discomfort, poor mental state and study status showed a trend of decreasing first and then increasing, and the lower limit temperature was about 29.5°C. However, when sitting, head discomfort decreased with decreasing floor temperature and was lower than standing. The lower limit temperature of the foot was 29.5 °C and 33°C for mental state and study status.

4 Conclusions and limitation

After the regression analysis of skin temperature, TSV, TPV and the incidence of sick symptoms, the main conclusions were as follows :

Different levels of temperature stratification appear on different floor surface temperatures, and the higher the temperature, the more obvious the sick phenomenon.The floor temperature directly affected the skin temperature and thermal comfort of the foot, and presented a statistically significant relationship with the overall and head thermal sensation.

According to the linear fitting and polynomial fitting, took the thermal sensation voting value of -0.5 to +0.5 for an acceptable range. It was calculated that the acceptable floor surface temperature of sitting was 26.1 °C ~ 34.3 °C, and 24.6 °C ~ 34.7°C for standing.Considering the local thermal comfort of the head and the long-term thermal exposure cumulative effect on health, this experiment suggested that the upper limit of the floor temperature should be 31 °C.

Only 12 subjects were involved in the experiment and the sample size of which is small, lead to a weak conclusion. Therefore, if the general applicability of national standards is taken into account, the sample size should be expanded in subsequent experiments to determine the upper and lower limits of the acceptable floor temperature range.

Acknowledgements

This research is supported by The UK-China collaborative research project "Low carbon climate-responsive Heating and Cooling of Cities (LoHCool)" supported by the National Natural Science Foundation of China (NSFC Grant No. 51561135002) and UK Engineering and Physical Sciences Research Council (EPSRC Grant No. EP/N009797/1), and the Fundamental Research Funds for the Central Universities (Grant No. 2018CDJDCH0015) and the 111 Project (No. B13041).

Abbreviations

Symbol	Definition	Unit/ Range
tw	Water supply temperature	°C
TSV	Thermal sensation vote	-3~+3
TPV	Thermal pleasure vote	-3~+3
MST	Mean skin temperature	°C
PD	Percentage dissatisfied	%
PS	Percentage satisfied	%

References

1. Bojic, Milorad, Cvetkovic, Dragan, Bojic, Ljubisa et al. Decreasing *energy use and influence to environment by radiant panel heating using different energy sources*. Applied energy, 2015, **138**(Jan. 15): 404-413
2. Ashrae. ANSI/ ASHRAE Standard 55-2013: *Thermal Environmental Conditions for Human* Occupancy. Atlanta, Ashrae, 2013.
3. ISO. EN ISO 7730:2005, *Ergonomics of the thermal environment - Analytical determination and interpretation of thermal comfort using calculation of the PMV and PPD indices and local thermal comfort criteria*. Geneva, Iso, 2005
4. ISO. PD ISO TS 13732-2: 2001 Ergonomics of the thermal environment-*Methods for the assessment of human responses to contact with surfaces. Human contact with surfaces at moderate temperature*. Geneva, Iso, 2001
5. ISO. EN ISO 13732-1:2006 Ergonomics of the thermal environment -*Methods for the assessment of human responses to contact with surfaces. Part 1: Hot surfaces* . Geneva, Iso, 2008
6. Olesen B W. *THERMAL COMFORT REQUIREMENTS FOR FLOORS OCCUPIED BY PEOPLE WITH BARE FEET*. Ashrae Transactions. 1977, **83**: 41-57.
7. Olesen B W, Mortensen E, Thorshauge J, et al. Thermal *comfort in a room heated by different methods*. Ashrae Transactions. 1980, **86**(1): 34-48.
8. Chongqing University, China Academy of Building Research. GB/T50785-2012 *"Evaluation Standards for Indoor Thermal and Humid Environment of Civil Buildings"* . Beijing: China Building Industry Press, 2012.
9. Sohn J Y. *THE STATE OF THERMAL SENSATION RESEARCHES IN KOREA AND THERMAL COMFORT IN ONDOL SPACE*. Human thermal environment report. 1986, **10**: 93-98
10. Song G S. *Buttock responses to contact with finishing materials over the ONDOL floor heating system in Korea*. Energy & Buildings. 2005, **37**(1): 65-75.
11. Song G S. *Effect of floor surface temperature on blood flow and skin temperature in the foot*. Indoor Air. 2008, 18(6): 511-520.
12. Zhang L, Emura K, Nakane Y. *A proposal of optimal floor surface temperature based on survey of literatures related to floor heating environment in Japan*. Applied Human Science. 1998 Mar;**17**(2):61-66.
13. Zhang H. *Human thermal sensation and comfort in transient and non-uniform thermal environment*. California: UC Berkeley, 2003.
14. Mishra A K, Loomans M G L C, Hensen J L M. *Thermal comfort of heterogeneous and dynamic indoor conditions -An overview*. Building & Environment. 2016, **109**: 82-100.
15. Gook-Sup Song. *Effect of floor surface temperature on blood flow and skin temperature in the foot*. Indoor Air. 2008;**18**:511-520.

16. Levick J. R, Michel C. C. *The effects of position and skin temperature on the capillary pressures in the fingers and toes*. The Journal of Physiology. 1978, **274**(1): 97-109.

17. Kräuchi Kurt. *Warm feet promote the rapid onset of sleep*. Nature. 1999, **401**:36-37

18. Sun Pi-Chang. *Relationship of skin temperature to sympathetic dysfunction in diabetic at-risk feet*. Diabetes Research and Clinical Practice. 2006, **73**(1):41-46.

19. Assanasen Paraya. *Warming of feet elevates nasal mucosal surface temperature and reduces the early response to nasal challenge with allergen*. Journal of Allergy and Clinical Immunology. 1999, **104**(2):285-293.

20. Liu H, Wu Y, Li B, et al. *Seasonal variation of thermal sensations in residential buildings in the Hot Summer and Cold Winter zone of China*. Energy & Buildings. 2017, **140**: 9-18.

21. Brager G S, De Dear R J. *Thermal adaptation in the built environment: a literature review*. Energy and Buildings. 1998, **27**(1): 83-96.

22. PARKINSON, T., DE DEAR, R. & CANDIDO, C. *Thermal pleasure in built environments: alliesthesia in different thermoregulatory zones*. Building Research and Information, 2016, **44**, 20-33.

23. Ramanathan N L. *A new weighting system for mean surface temperature of the human body*. Journal of Applied Physiology. 1964, **19**(3): 531-533.

24. GAGGA, A P.; STOLWIJK, A J. , and HARDY, J D. *Comfort and thermal sensations and associated physiological reponses at various ambient temperatures*.Environment Research. 1967,**1**:1-20.

Simulation of CO and CO_2 emissions in model combustion chamber based on the combination LES and Reactor Network model

Matveev S.S. , *Idrisov* D.V, *Gurakov* N.I, *Gopalakrishna* Gangisetty, *Zubrilin* I.A., *Matveev* S.G *and Shchepakina* E.A

Scientific educational center of gas-dynamics research, Samara National Research University, Samara City, Russia

Abstract. Air pollution is a major concern of recent decades, which has a serious toxicological impact on human health and the environment. It has a number of different emission sources, but one of the main sources of environmental pollutions is transported systems, in particular aviation gas turbine engines (GTE). Currently environmental issues of GTE are mainly solved by using semi-empirical techniques and experimental refinement of prototypes. In this paper we presented an algorithm for simulating the emission of CO and CO_2 in a model combustion chamber under various initial conditions and compared the results, validated with an experimental data.

1 Introduction

Air pollution is a major concern of recent decades, which has a serious toxicological impact on human health and the environment. It has a number of different emission sources, but one of the main sources of environmental pollution is transported systems in particular, aviation gas turbine engines (GTE). The International Civil Aviation Organization (ICAO) established standards for limiting the emissions of nitrogen oxides (NOx), carbon emissions such as carbon monoxide (CO), carbon dioxide (CO_2), unburned hydrocarbons (UHC) and smoke. At the same time, it is planned to gradually strict the norms, including 60% of NOX emissions reduction by the year 2030 (compared to 2008 norms) [1].

The combustion chamber (CC) is the most critical part of the gas turbine engine (GTE), which determines its ecological characteristics. Currently, the environmental issues of GTE are mainly solved by using semi-empirical techniques and experimental refinement of prototypes. This approach is characterized by increased time-consuming and low informative processes in combustion chamber, which does not allow ensuring the implementation of promising ICAO standards.

Perspective methods for calculating the characteristics of turbulent combustion of hydrocarbon fuels are based on the joint solution of non-stationary equations of gas dynamics and detailed chemical kinetics by direct numerical simulation (DNS) methods. However, at the present time, in such a formulation, the solution of problems with respect to combustion turbojets cannot be realized because of the limited possibilities of advanced computer technology. The solution to this problem is seen in the application of combined methods, consisting in a sufficiently accurate solution of each subtask separately and combining them into a single algorithm. Simulation of a chemical kinetics is impossible without the use of detailed and reduced reaction mechanisms, for the formation of which it is necessary to know the exact composition of the initial fuel.

2 Experimental setup

This experiment was carried out with a single-burner modeled combustion chamber (Figure 1). The model of the combustion chamber is fixed by means of a flange (2) to the experimental installation and includes the following main elements: diffuser (1); the combustion chamber body (3); burner device (4); the body of the flame tube (5); output device (6). One of the side walls (9, 10) of the combustion chamber has a window with heat-resistant glasses (7, 8), for the possibility of visualizing the processes. The experiment was carried out at α = 1.15 ... 1.67 ($\alpha=1/\varphi$), initial temperature + 150 ° C and atmospheric pressure. The pressure ratio in the combustion chamber is constant maintained at ΔP = 3%.

$$\Delta P = \frac{P_{in} - P_{out}}{P_{in}} * 100\%$$

P_{in} – total pressure at inlet of combustion chamber (const);

P_{out} - total pressure at outlet of combustion chamber (const).

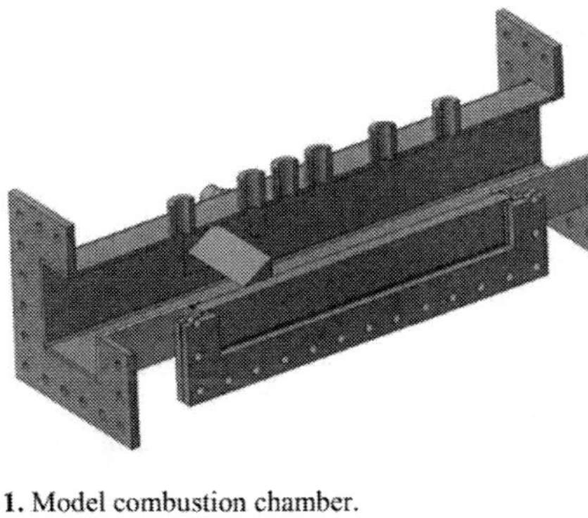

Fig. 1. Model combustion chamber.

Experimental trials were carried out at the high-temperature experimental setup with a model combustion chamber, which was described in the paper [2].The general view of the high-temperature installation connected to the fuel line and air passage lines is shown in Figure 2.

1- the main air heater N = 102 Kw; 2 - two-circuit nozzle for supplying liquid and gaseous fuels; 3 - a section for preliminary preparation of the fuel-air mixture; 4 - diffuser; 5-device for equalizing the flow; 6-measuring section; 7-model of the combustion chamber; 8 - output section; 9 - an exhaust shaft; 10 - ignition system power module; 11 - gas pressure regulator.

Fig. 2. Experimental installation with a model combustion chamber

The sample line consists of a sampler (1), a Richter absorber (2); a pump (3) built into the SICKGMS–810 analyzer (4), Seger pipettes (5), and a GSB-400 gas meter (6). GSB-400 is used to estimate the volume (and the flow rate of sampled gases. The sample from the sampling point to the Seger pipettes - was pumped with a pump built into the SICKGMS–810 analyzer.This sample line configuration allowed simultaneous sample pumping via Seger pipettes, its dehydration and analysis. During sampling, the combustion products were pumped via the Seger pipettes at the flow rate of the volume equal to 20 pipette volumes. Obtained gas samples were analyzed using gas chromatography. The sampling was carried out at the output of the simulated combustion chamber. Statistical analysis of measurement errors 0.95. The error of a single measurement is ± 10%, the error in determining the mean values from the results of five experiments is ± 4%.

3 Simulation

The geometric model of the combustion chamber with an angular flame stabilizer is made like experimental model and is shown in Figure 3.

Since the geometric model of the CC has rectangular forms of the working area, in this study a structured mesh was calculated successfully, generated in an automatic mode with a total number of elements of the order of 0.4 million pieces. Since intense vortex flows are formed in the zone behind the flame stabilizer [3,4], the size of the finite elements in the corresponding regions of the CC has been reduced. The total number of items increased to 0.6 million pieces. The final calculated grid model is shown in Figure 4

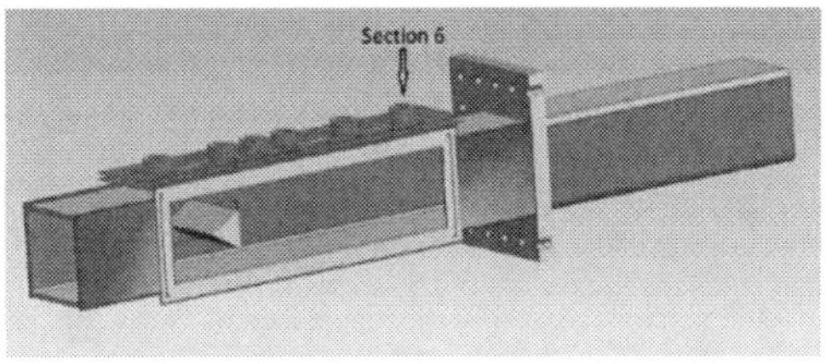

Fig. 3. Geometrical model of the combustion chamber with a corner flame regulator

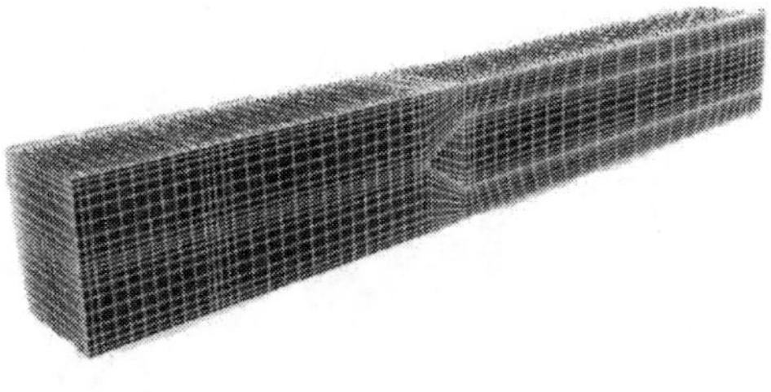

Fig.4. Grid model of the angular flame stabilizer

4 Laminar flame speed

One of the quantities characterizing the combustion process in the combustion chamber is the normal flame propagation speed S_L, which characterizes the velocity of the flame front relative to the unburned gas in a direction perpendicular to its surface. It is a fundamental characteristic of the fuel-air mixture, depending on the fuel surplus, initial temperature and pressure:

$$S_L = f(\phi, T, p, \text{fuel type})$$

Due to the fact in the standard settings of ANSYS Fluent S_L is set by a constant for the entire range of temperature and pressure values, and the results with a high error. Therefore, when determining SL in the mathematical model of calculation, it is necessary to know its dependence on temperature and pressure. In [5], such function was obtained on the basis of numerical simulation results in the software product ANSYS ChemKin-Pro, using the kinetic mechanism of GRI 3.0 [6].

5 RANS/LES

Modes of operation of the model combustion chamber varied over a wide range in terms of the coefficient of excess air, residence time and the initial temperature of the fuel assembly. The temperature (Tc) varied from 373 to 523 K; the excess air factor (α) varied from 1.1 to 1.7; the speed at the inlet (u) varied from 3 m/s to 15 m/s. The pressure drop in the model CC with the angular flame stabilizer was ΔP* = 3% for the velocity u = 10 m/s.

The combustion of the pre-prepared fuel assembly was realized using the flamelet model. At the first stage, calculations were made in the RANS setup, which was then used to simulate the large-eddy simulation (LES) [7]. The main investigated mode of operation of the model combustion chamber was that, regime with the following initial conditions: $\alpha = 1.36$, $T_k = 423K$. The results of calculations of the concentrations of the final combustion products of CO_2 and H_2O along the axis of the model CC, as well as their comparison with the experimental data are shown in Figures 5(a) and 5(b).

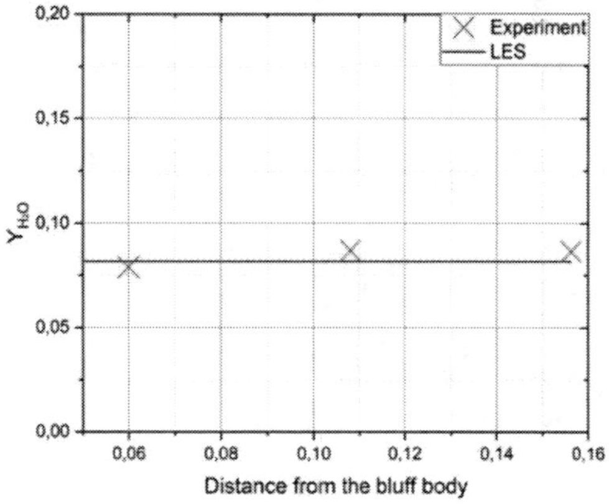

Fig.5a. concentration of H_2O

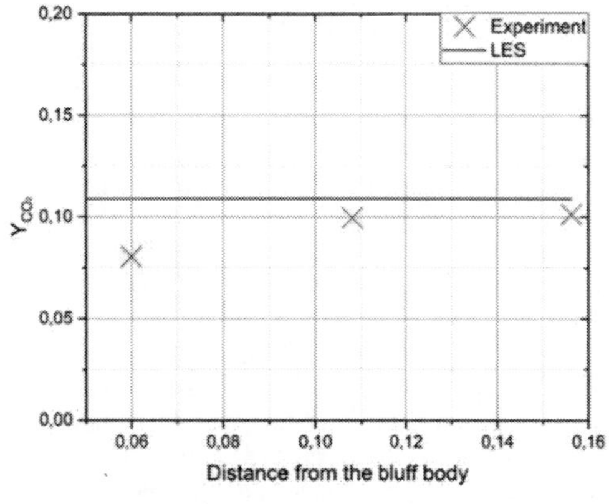

Fig.5b. concentration of CO_2

The results of the calculation showed a satisfactory agreement with the experimental data. The correct prediction of the final combustion products of the fuel guarantees an adequate temperature distribution, which is necessary for the formation of the reactor chain.

6 Reactor network

The reactor circuit is used to simulate the combustion products in the combustion chamber with the possibility of using detailed kinetic mechanisms, in particular emissions of pollutants such as NOx, CO and unburned hydrocarbons. The combustion chamber volume is automatically broken down into ideally stirred reactors. Since the set number of reactors is much smaller than the size of the finite elemental mesh of the CFD model, the reactor models allow significantly accelerating the modeling of concentration fields. As a rule, simulation with the help of reactors is performed on a convergent stationary RANS solution or a time-averaged non stationary LES solution [7].

7 Results

Based on the calculation procedure which is described above, a study was made for the effect of the excess air factor (α), velocity and temperature of the inlet stream in the CC on the formation of carbon oxides (CO and CO_2).

The Figures 6.a, 6.b and 7.a, 7.b shows data on the effect of the air-fuel mixture composition, while the velocity and temperature of the flow at the inlet to the model CC were kept constant: u = 10 m/s; Tk = 423 K. The results of calculations in the Section no.6 of the model combustion chamber were not averaged over the whole plane, but only on the surface in the central section measuring 16x16 mm. precisely on this section 6, averaging of the experimental data obtained with the aid of the probe method of sampling was performed.

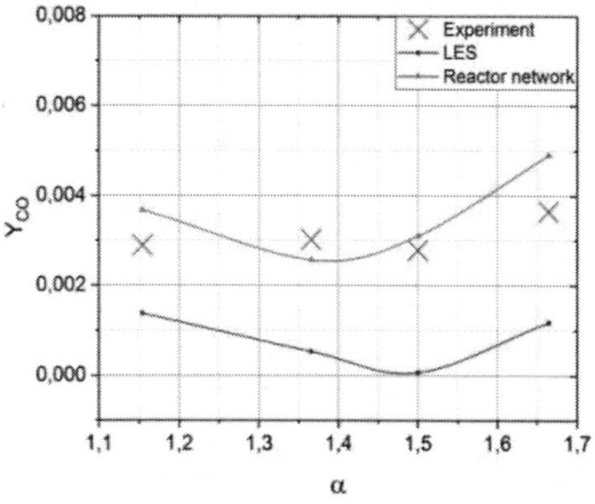

Fig.6a. Mass concentration of CO as a function of α

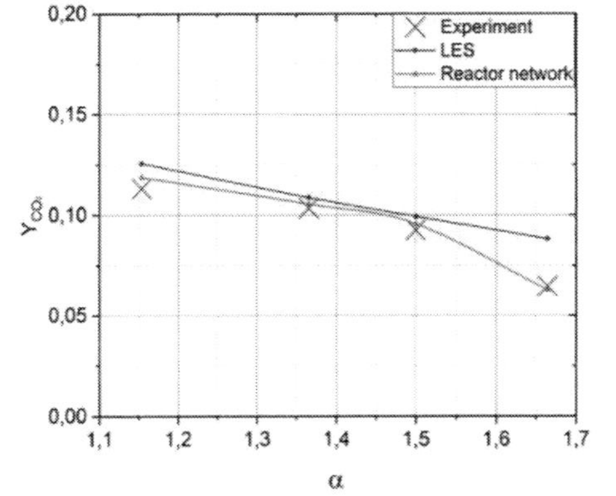

Fig.6b. Mass concentration of CO_2 as a function of α

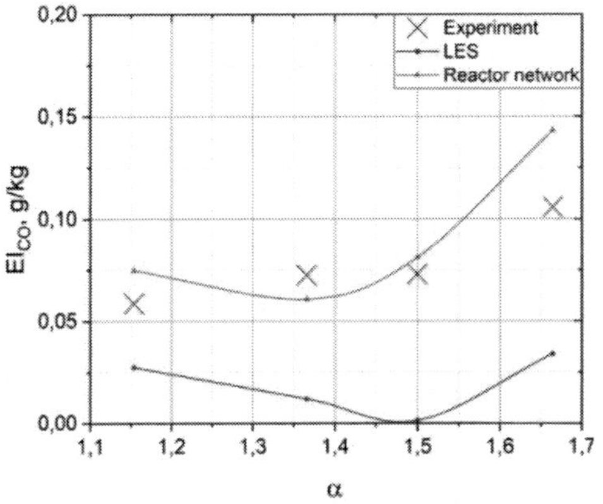

Fig.7a. CO emissions as a function of α

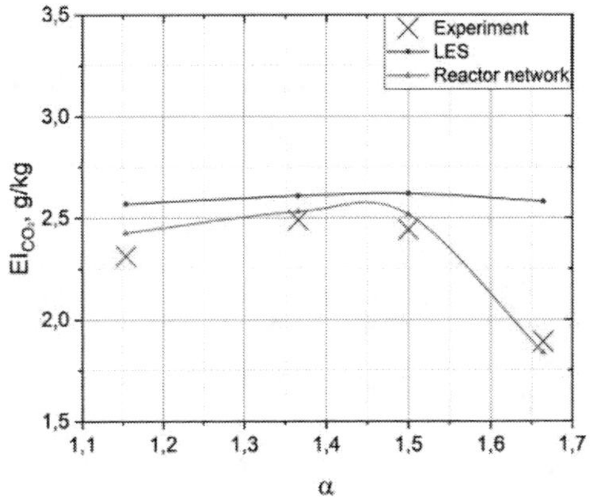

Fig.7b. CO_2 emissions as a function of α

From the experimental data shown in the Fig. 6.a and Fig. 6.b, it is seen that with an increase in α from 1.15 to 1.67, the CO emission index increases from 0.06 to 0.11 g/kg of fuel, which is associated with a deterioration in the combustion conditions of depleted pre-prepared fuel-air mixtures and corresponds to well-known tendencies. The results of calculations by the LES approach (without simulating the chain of reactors) give significantly lower levels of CO concentrations: EI_{CO} = 0.01 ... 0.03 g / kg of fuel, which is 4-5 times less than the experimental data.

The use of reactor models allows a better qualitative prediction of the emission of both CO and the final product CO_2. The deviation from the experimental data on carbon monoxide (CO) for α = 1.15 ... 1.5 is not more than 15%, and for α = 1.67 not more than 25%.

The influence of the flow velocity at the inlet to the model combustion chamber with the angular flame stabilizer is shown in Fig. 8.a and Fig.8.b, while the inlet temperature in the model CC and the excess air coefficient were kept constant: α = 1.36; T_K = 423 K.

An increase in the flow rate leads to a decrease in the residence time of the fuel assembly in the combustion zone behind the corner flame stabilizer. This estimation shows that increasing the flow rate from 3 to 15 m/s leads to a decrease in the residence time in the zone from the corner flame stabilizer to the combustion chamber section No. 6 from 20ms to 4ms. This leads to an increase in the concentration of carbon monoxide and, accordingly, to a decrease in the final product of combustion of CO_2. This trend is not reflected in the modeling of the LES approach, while the connection of the reactor circuits allows to obtain more reliable results with a maximum deviation from the experimental data within the limits up to 10%.

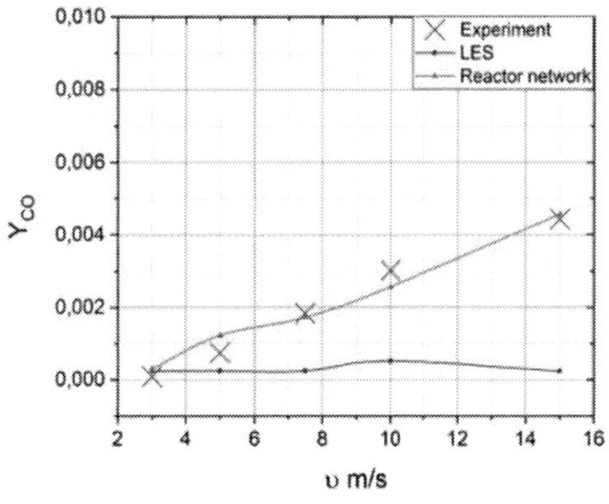

Fig.8a. Mass concentration of CO depending on initial flow rate

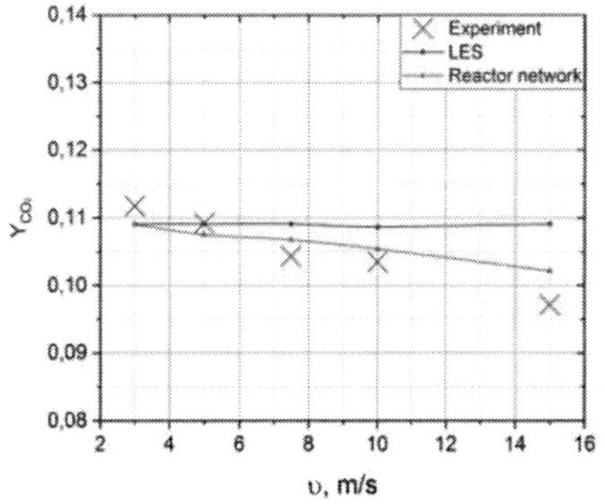

Fig.8b. Mass concentration of CO_2 depending on initial flow rate

Fig.9.a and Fig.9.b shows the effect of the initial temperature of the pre-prepared fuel-air mixture on the

formation of CO and CO_2, while the flow rate and the composition of the fuel assemblies were kept constant:
u = 10 m / s; α = 1.36.

With an increase in the initial temperature, previously prepared by FA, the processes of ignition and combustion are improved, which is reflected in the experimental data obtained in this paper. An increase in temperature from 373 to 523 K leads to a decrease in the CO concentration from 0.0036 to 0.0018. Simulation using a chain of reactors as a whole with satisfactory accuracy (up to 15%) reflects the experimental data, and direct LES simulation produced significantly lower results.

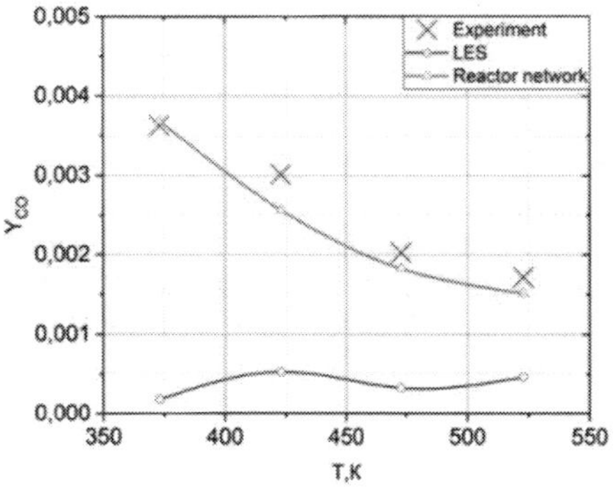

Fig.9a. Mass concentration of CO as a function of the initial flow temperature

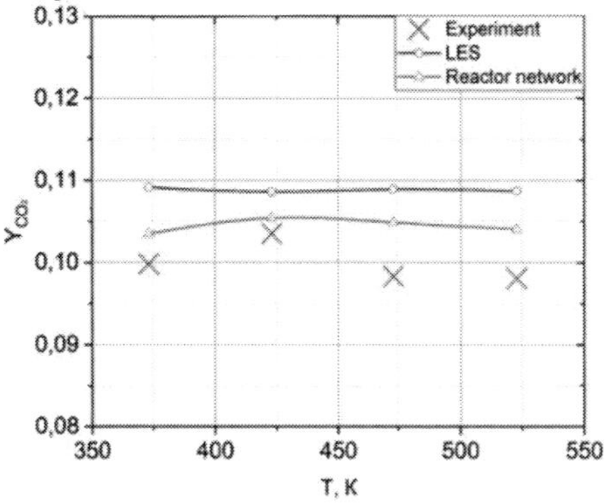

Fig.9b. Mass concentration of CO_2 as a function of the initial flow temperature

8 Conclusions

As a result of computational and experimental studies, combustion of natural gas in a model combustion chamber with an angular flame stabilizer was studied. Mathematical models of turbulence, combustion, environmental properties, and initial and boundary conditions are determined and described. The kinetic mechanism of methane oxidation GRI 3.0 was experimentally confirmed using the example of calculating the normal flame propagation velocity, and the possibility of its application to natural gas was also proved.

The algorithm for determining CO and CO_2 has been formulated and approved, in which calculations are performed in the RANS at the beginning, on the basis of which the calculation in the LES is performed. The results obtained by the LES approach show satisfactory accuracy in predicting the final products of the fuel oxidation reaction (CO_2 and H_2O), as well as the temperature distribution. The chain of reactors for combustion of the pre-prepared fuel-air mixture is formed on the basis of the temperature field obtained using the LES approach. Modeling using reactor models uses the detailed kinetic mechanism of fuel oxidation, verified with experimental results of the normal flame propagation velocity. Modeling of CO_2 and CO by this algorithm showed good agreement with the experimental results.

Acknowledgments

This work was supported by the Ministry of education and science of the Russian Federation in the framework of the implementation of the Program "Research and development on priority directions of scientific-technological complex of Russia for 2014-2020" (RFMEFI58716X0033)

References

1. International standards and recommended practice "Environnemental protection".Annex 16 to the Convention on International GA, Volume 2, "Emission of aircraft engines" [Text] / Montreal, 2nd ed.1993- 96 p.
2. Matveev, S.S., Chechet, I.V., Semenikhin, A.S., Abrashkin, V.Y., Lukachev, S.V., Matveev, S.G. Experimental Study of the Combustion of Kerosene and Binary Surrogate in the Model Combustion Chamber. Journal of Combustion, Volume 2017, Article ID 3963075, 6 pages.
3. Matveev, S.G. Modeling of the flow structure behind the flame stabilizer by the method of large eddies [Text] / S.G. Matveev, I.A. Zubrilin // Proceedings of the Samara Scientific Center of the Russian Academy of Sciences, - Т. 15, - № 6-4, 2013, - p. 874-880.
4. Matveev, S.S., Zubrilin, I.A., Orlov, M.Y., Matveev, S.G., Chechet, I.V. Investigation of fuel distribution in partially premixed swirled burner with pilot flame. Proceedings of the ASME Turbo Expo 4B-2016, GT2016-57478
5. Zubrilin, I.A., Matveev, S.S.,Matveev, S.G.,Idrisov, D.V. Measurements and Experimental Database Review for Laminar Flame Speed Premixed Ch4/Air Flames IOP Conference Series: Materials Science and Engineering, Volume 302, Issue 1, 8 February 2018, № 012078
6. GRI-Mech 3.0, The Gas Research Institute, http://www.me.berkeley.edu/gri_mech/
7. ANSYS 18 User Guide Inc. http://www.ansys.com/

https://doi.org/10.1051/e3sconf/20198003009

Heavy metal contamination in bottom sediments of saline endorheic lakes of Kulunda and Barabinsk steppes (Western Siberia, Russia)

Marina Kolpakova

Institute of geology and mineralogy SB RAS, 634000 Novosibirsk, Russia

Abstract. The article presents the results of assessment of heavy metals contamination in bottom sediments of saline lakes located on the territory of Kulunda and Barabinsk steppes of Western Siberia, Russia. The territory is characterized by large amount of lakes (more than 5000), including the Chany lake, which is the biggest lake of south of Western Siberia. By calculation the geochemical parameters, such as Enrichment factor, Contamination factor, Geoaccumulation index, Pollution load index, the common level of bottom sediments was classified as unpolluted. However, in comparison with average level of bottom sediment on that territory the lake sediments are enriched with the following elements: Sr> Rb> Zn> V> Cr> Ni> Cu> As> Pb> Th> U> Mn> Fe> K> Ti.

1 Introduction

According to [1], chemical elements from natural and anthropogenic sources can flow into aquatic ecosystems directly or through the atmosphere (for example, as a result of agro-economic work), while creating certain geochemical conditions of the environment that can accumulate or remove heavy metals from the system. Particular attention in connection with the pollution of heavy metals requires water bodies, the catchment area of which includes industrialized areas.

The main polluters in the southern part of Siberia are the enterprises of chemistry, petrochemistry, power engineering, municipal services and processing of agricultural products [2, 3]. For example, in the Altai Territory, with the exception of two landfills for disposal (Altaihimprom OJSC (Yarovoye) and Slavgorod radio plant), there are no specially created sites for industrial and household waste, despite the fact that each year adds an edge about 400 thousand tons of household waste and 750 thousand tons of industrial waste. In addition, out of 1600 villages, only 20 have sewers with sewage treatment facilities [3].

The main substances that cause pollution of surface water in recent years remain sulfates, chlorides, general phosphorus, ammonium nitrogen, nitrates, fats, petroleum products, general iron and phenols volatile. In addition, it is possible to add chemical oxygen consumption 2-3 times higher than maximum permissible standards.

When it comes to lake ecosystems, which act as collectors of all these discharges and substances, the scale of the problem voiced within the projects becomes evident. Details of sources and activities in the territories are presented in the article.

2 Regional overviews

The research presented in this paper were carried out at the territory of Kulunda and Barabinsk Steppes (Altay and Novosibirsk regions, Russia) where more than 5000 saline lakes with total area about 4000 km^2 are located [2]. The studied area forms an almost flat basin with a predominantly steppe landscape. The map of studied lakes presented in Fig. 1.

Fig. 1. Map-scheme of studied lakes

The predominant anion in most lakes in the western part of the Kulunda steppe is a chloride ion (concentration range from 0.2 to 190 g/L). The content of carbonate components ranges from 0.5 to 4.3 g/L: in soda lakes type its content reaches an average of 25 eq.% while its concentration in chloride lakes is only up to 0.8 eq.%. The concentration of sulfate ions varies in different lakes from 0.09 to 65 g /L. The carbonates dominate in lakes with relatively low mineralized water and if the value of TDS is higher than 41.1 g/L and pH more than 9.0. With the increasing of salinity and the decreasing of pH the anionic composition changes from HCO3-Cl-SO4 to Cl-SO4-HCO3, what is more the increase in the chloride content is significantly ahead of the growth of sulfate. This is possible in the case of water saturation with respect to sulphate minerals (gypsum, thenardite, mirabilite, etc.) and their precipitation, as well as lack in accumulation of sulfates can be a result of its bacterial reduction.

Based on chemical composition, all studied lakes were divided into two large groups.

Water Type 1 (Cl-Na) chloride lakes with Cl-ions in dominance. Chloride lakes are the most widespread type (50%) among all studied lakes. Value of TDS varies between 40 and 600 g/L, pH between 7.2 and 9.4. This is the most saline type of lakes and the least alkaline in Kulunda steppe. Malinovoe lake – the most saline lake of Kulunda steppe – is included in this type (highest TDS is 560 g/L (2011), pH is 8.4 (2014)). Among the cations, sodium prevails (average is 74 eq.%), magnesium in waters of this type is somewhat larger than in soda and is equal to 26 eq.%.

Water type 2 (Cl-Na with HCO3 up to 30 eq.%) chloride lakes with high soda amount (Cl>50%, HCO3+CO3 up to 30%) Even a small amount of soda in water can provide high pH values. As can be seen from the picture, these two types have the similarities in major ions distribution. The difference is only in CO3, SO4, Cl and Mg accumulation.

The trace element composition of the lake waters of the Altai Territory is of undoubted interest from the point of view of their hydromineral resources. Among the trace elements in lake waters, elevated boron contents (up to 78 mg/L), bromine (up to 820 mg/L), lithium (up to 3.6 mg/L), strontium (up to 13 mg /L), iodine up to 6.3 mg /L), arsenic (up to 0.6 mg/L).

3 Methods and materials

The major components of water samples were analyzed at the "Voda" Accredited Center of Tomsk Polytechnical University (Tomsk) following the methods described in the literature. The contents of trace elements were analyzed on an Agilent 7500a ICP MS at the Institute of Solid State Chemistry and Mechanochemistry, Siberian Branch, Russian Academy of Sciences; Li was determined using a Varian AA 280 FS spectrometer.

In order to estimate the level of accumulation of chemical elements in bottom sediments of lakes several geochemical parameters including following indexes.

Enrichment factor (EF) was calculated as a ratio of element normalized concentrations between studied sample and reference data [4] by following equation [5]:

$$EF = \frac{\left(\frac{C_n}{Fe}\right)_{sediment}}{\left(\frac{C_n}{Fe}\right)_{reference}}, \quad (1)$$

where Cn is a concentration of chemical element n in bottom sediments. All calculations were made in comparison with geochemical background data for Western Siberian Platform [4]. The normalization of chemical elements was conducted on iron content. Assessment criteria for EF suggested by [5]: indicated no enrichment (EF<1), minor enrichment (1<EF<3), moderate enrichment (3<EF<5), significant enrichment (5<EF<10), severe enrichment (10<EF<25), very severe enrichment (25<EF<50), and extremely severe enrichment (EF<50).

Geoaccumulation index (Igeo) was calculated as [6]:

$$I_{geo} = log_2\left[{C_n}/{(1{,}5 \times B_n)}\right], \quad (2)$$

where B_n is an average element concentration [4]. Seven classes of Igeo are determined by [6]: unpolluted (0≤Igeo), unpolluted to moderately polluted (0≤Igeo≤1), moderately polluted (1≤Igeo≤2), moderate to strongly polluted (2≤Igeo≤3), strongly polluted (3≤Igeo≤4), strongly to extremely polluted (4≤Igeo≤5), and extremely polluted (Igeo≥5).

The level of element contamination was estimated by the contamination factor (CF) [7] as a ratio between the metal content in the sediment sample and background values [4]:

$$CF = {C_{sample}}/{C_{background}}, \quad (3)$$

where CF<1 indicates low contamination; 1<CF<3 is moderate contamination; 3<CF<6 is considerable contamination; and CF>6 is very high contamination

Contamination degree (Cd) of bottom sediments was defined as a sum of all [8]: low contamination degree (Cd<6), moderate contamination degree (6≤Cd<12); considerable contamination degree (12≤Cd<24); very high contamination degree (Cd≥24).

Modified degree of contamination (mCd) was estimated according to the formula [9]:

$$mCd = \frac{\sum_{i=1}^{i=n} CF}{n}, \quad (4).$$

The Pollution load index (PLI) [10] represents the number of times by which the heavy metal concentrations in the sediment exceeded the background concentration and give a summative indication of the overall level of heavy metal toxicity in a particular sample and is determined as the nth root of the product of nCF:

$$PLI = (CF_1 \times CF_2 \times CF_3 \times \ldots.. \times CF_n)^{1/n}, \quad (5),$$

 https://doi.org/10.1051/e3sconf/20198003009

where CF_n is the CF value of element n. The PLI values are interpreted into two levels as polluted (PLI>1) and unpolluted (PLI<1).

The calculation of all indices was carried out with the normalization to background concentrations since the concentrations of many elements in the territory of Western Siberia are higher than their values for the upper continental crust.

4 Results and discussion

An assessment of the geological and geochemical factors affecting the formation of the geochemical balance of lakes has shown that the main sources of entry of elements into saline lakes are the underground and sewage waters of the drainage basins.

Fig. 2 indicates that enrichment factor for Cu is the highest between all heavy metals and varies from 2 to 9, indicating that there is an anthropogenic source for that element. High enrichment for Sr is explained by salinization processes and formation of secondary minerals (Strontianite, Celestite). All other elements Ni, As, Zn, Th, U show low level of enrichment (maximum 2-6) that shows the natural sources for them.

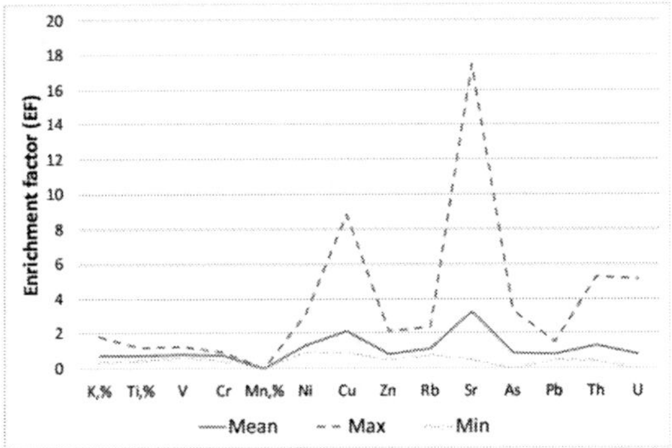

Fig. 2. Enrichment factor (EF) for chemical elements in bottom sediments of saline lakes of the south of Western Siberia

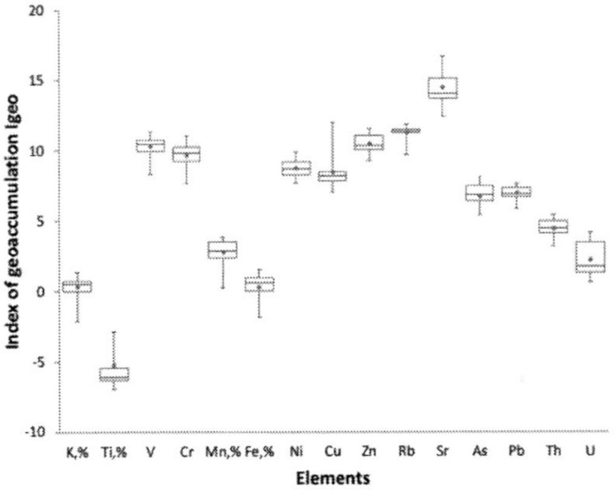

Fig. 3. Box-and-whisker plot of the geoaccumulation index of studied metals in 15 surface sediment samples of saline lakes

The estimation of geochemical indices of contamination of bottom sediments of lakes has shown that, based on the average values of the geoaccumulation index, the lake sediments are enriched with the following elements: Sr> Rb> Zn> V> Cr> Ni> Cu> As> Pb> Th> U> Mn> Fe> K> Ti (Fig. 3).

Table 1. Contamination factors (C_f), contamination degree (C_d) and Pollution load index (PLI) for studied lakes

Lake ID	Contamination degree		Pollution load index (PLI)
	Classic (Cd)	Modified (Cp)	
Petuhovo	13.07	0.87	0.28
Petuhovo-2	11.45	0.76	0.37
Kulundinskoe	7.71	0.55	0.26
Kuchukskoe	3.03	0.23	0.09
Burlinskoe	4.87	0.35	0.17
Saline	11.80	0.84	0.39
Saline-2	11.25	0.75	0.39
Kamenskoe	7.39	0.53	0.28
Small Bagan	7.90	0.53	0.28
Ostrovnoe	8.58	0.61	0.32
Celebnoe	6.99	0.50	0.26
Chany	12.61	0.84	0.45
Sartlan	23.50	**1.68**	0.53
Yarkul'	13.89	0.99	0.45
Yudinskoe	16.85	**1.30**	0.56

Moreover, the extreme pollution level for this index is observed for the most part of the elements, including for Rb, Zn, V, Cr, Ni, Cu, As, Pb. Despite this, most of the lakes remain in the category of non-contaminated sites, with the exception of objects on the territory of which fishing, industrial or tourist activities are carried out (Chany, Sartlan, Yarkul Lakes, Table 1). A significant level of pollution is manifested mainly for Cu, Ni, Fe, U.

5 Conclusion

The results obtained in the framework of the project can be used by the authorities to manage the environment. The flow of various wastewater (industrial, economic, agricultural) into the lakes of Western Siberia, including underground and land flows, is the main source of pollution of lake water and precipitation. As preliminary recommendations, the authors of the project believe that the following approaches can be applied to preserve and prevent environmental pollution: 1. Develop a monitoring network to initiate preventive actions as part of the annual observations of the elemental composition of heavy metals in waters and bottom sediments. 2. Reduction of anthropogenic emissions. To this end, the ecological rehabilitation of the river waters of the region and the conduct of high-quality treatment of sewage entering them. 3. Prevention of direct entry of agricultural wastewater. 4. Minimizing the impact of the mobilization of metals in the environment by improving the quality of water through the extraction of highly polluted precipitation.

The data assessment and study were founded by Russian Science Foundation (RSF, Project 17-77-10086)

References

1. Holland HD, Turekian K. *Geochemistry of Earth Surface Systems* (Elsevier, 2010)
2. Borzenko SV, Kolpakova MN, Shvartsev SL, Isupov VP. Biogeochemical conversion of sulfur species in saline lakes of steppe Altai. Journal of Oceanology and Limnology, **36**(3), 676-686 (2018)
3. Isupov VP, Kolpakova MN, Borzenko SV, Shatskaja SS, Shvartsev SL, Dolgushin AP, Lyakhov NZ. Uranium in the mineralized lakes of Altai Krai. Doklady Earth Sciences, **470**(2), 1067-1070 (2016)
4. Strakhovenko VD, Shcherbov BL, Malikova IN, Vosel' Y. The regularities of distribution of radionuclides and reare-earth elements in bottom sediments of siberian lakes. Russian Geology and Geophysics, **51**(11), 1167-1178 (2010)
5. Birth GA. *A scheme for assessing human impacts on coastal aquatic environments using sediments.* (Woodcoffe CD, Furness RA, Eds. Coastal GIS 2003)
6. Müller. Index of geo-accumulation in sediments of the Rhine River, Geo. J., **2**,108-118 (1969)
7. Turekian KK, Wedepohl KH. Distribution of the elements in some major units of the Earth's crust, Geol Soc Am Bull **72**: 175-192 (1961)
8. Hakanson L. An ecological risk index for aquatic pollution control.a sedimentological approach. Water Research, **14**(8), 975-1001 (1980)
9. Abrahim GMS, Parker RJ. Assessment of heavy metal enrichment factors and the degree of contamination in marine sediments from Tamaki Estuary, Auckland, New Zealand, Environ. Monit. Assess **136**: 227-238 (2008)
10. Harikumar PS, Nasir UP, MujeebuRahman MP Distribution of heavy metals in the core sediments of a tropical wetland system, Int J Environ Sci Technol **6**: 225-232 (2009)

Ecological aspects of economical development: issues of forecast greenhouse gas emissions in road transport in Europe and regions of Russia

Leontiy Eder[1, 2], *Irina* Filimonova[1, 2], *Vasily* Nemov[1, 2], *Anna* Komarova[1], and *Kirill* Sablin[2]

[1] Novosibirsk State University, Russia
[2] Federal Research Center of Coal and Coal Chemistry SB RAS, Russia

Abstract. Environmental aspects are an essential part of economic development. Improvement of the environmental situation can have a significant impact on the pace and structure of economic development. For environmental component, it is important to analyze the current state and predict greenhouse gas emissions. The development of methodological approaches in this area will allow for more detailed forecasting of the situation. In order to reduce the emission of greenhouse gases, European countries have set targets for efficiency of energy consumption and widespread use of renewable energy sources, which they have achieved and become world leaders in using them. By 2020, the goal is set for energy consumption to be at least 20% from renewable sources. According to the forecast on average, in 2040 the share of oil products in the structure of fuel consumption in road transport in Europe will be reduced to 80%. Nevertheless, in the countries of Europe, various trends in the field of greenhouse gas emissions are expected. Most countries have a high potential for reducing greenhouse gas emissions from fuel combustion on road vehicles. In Russia, emissions from vehicles are projected to reduce by 8% by 2040.

1 Introduction

Analysis of the current situation and forecasting in the energy sector is a key component in the formation of long-term policies for sustainable development of the economy [1, 2]. An important aspect is the relationship between the economy and its major industries, energy and environment. Transport is one of the largest and strategically important industries. Given that transport is the main consumer of petroleum products, the transport sector accounts for about 30 % of total energy consumption.

In order to reduce the emission of greenhouse gases, European countries have set targets for efficiency of energy consumption and widespread use of renewable energy sources, which they have achieved and become world leaders in using them.

By 2020, the goal is set for energy consumption to be at least 20% from renewable sources (Directive 2009/28/EC) and at least 27% by 2030 (2030 Energy Strategy). Forecast of energy consumption by vehicles in Europe and in Russia and its regions is needed to determine the long-term guidelines for the development of energy markets, development of transport infrastructure, petrochemical and related industries, ecology and environment.

At the same time, the rapid growth of greenhouse gas emissions in transport makes it the largest source of emissions, and the high growth rates of emissions over the past five years brought it to the pre-crisis level of 2008.

Motor transport is the largest and most variable source of pollution relative to other modes of transport. On one hand the volume of emissions is affected by the growth of the absolute number of vehicles, on the other hand, the ecological characteristics of traditional engines are improving, and hybrid technologies and electric motors with zero emission of carbon dioxide are being rapidly implemented.

The object of the study is the motor transport of the countries of Europe and regions of Russia. The subject of the study is methodological approaches to forecasting of the energy consumption by road transport and greenhouse gas emissions.

The purpose of the study is to establish long-term trends and forecasts for the energy consumption by road transport and estimation of greenhouse gas emissions.

The objectives of the study include: 1) identification of sustainable trends in the dynamics of motorization and structures of energy use in the road transport; 2) development of methodological approaches to the forecasting of the energy consumption by road transport; and 3) estimation of greenhouse gas emissions by fuel combustion in cars.

Scientific novelty of research consists of:

– proposed directions for improvement of the methodological approach to forecasting energy consumption in the transport sector, taking into account specific characteristics of the countries and regions and prerequisites for stable trends in improving the efficiency of energy consumption by road transport;

– multi-factor model built on the basis of established and systematized indicators, allowing to calculate the transport saturation level for the economy in the medium and long term perspective, depending on a number of selected factors;

– estimation of the greenhouse gas emissions in European countries and regions of Russia for the period up to 2040 on the basis of the forecasted number of vehicles, energy consumption and energy structure.

The paper is organized into following sections: (1) an introduction to the study; (2) a literature review; (3) the data and methodology; (4) results and discussion; and (5) conclusions.

2 Literature review

Both global and Russia's economy show persistent trends that differently influence energy consumption and structure of the road transport sector, including: 1) increased number of vehicles; 2) higher engine efficiency; 3) appreciable shifts in the structure of energy consumption.

In order to determine the long-term targets of energy consumption in the transport sector of Russia it is necessary to consider global trends, especially in the EU and to determine the possibilities of their application with the specificity of Russian regions taken into account.

2.1 Vehicle quantity dynamics

In most developed countries, the optimal vehicle saturation level per 1,000 population has been achieved. Given that energy consumption dynamics in the transport sector is most affected by increasing engine efficiency, this results in gradual reduction of energy consumption [3]. However, in most countries with developing economies, the current level of vehicle ownership is far from saturation, while the transport fleet growth rate is consistently high. As a result, despite the developing energy-saving technologies, energy consumption in transport sector is generally increasing [4, 5].

The relationship between the number of cars per capita and GDP is not linear. The specific number of vehicles is growing relatively slow at low levels of this indicator, which is followed by about two-fold increase in an average level (from $ 3,000 to $ 13,000 per capita) [6]. With further increase in GDP per capita, there is a slowdown in growth in the number of vehicles as long as the number of cars per capita reaches a certain level, which is interpreted to be vehicle saturation level in the economy [7]. This dependence is shown in Figure 1. Notably, we used the data from countries with different income levels during the period of 1991-2013.

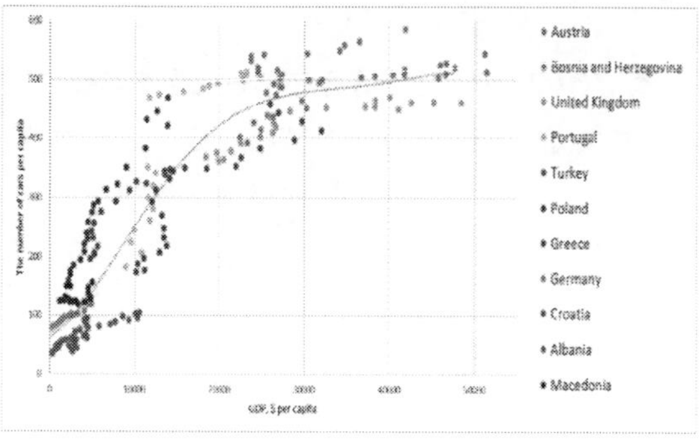

Fig 1. The relationship between number of vehicles and GDP per capita in some countries, according 1991-2013.Source: European Commission: Eurostat; World

As a result, the relationship between the number of vehicles and the GDP per capita can be represented by an S-shaped curve. There are a lot of different functional forms that can describe such process, for example, logistics, and logistics logarithmic, normal storage. Gompertz function shows the most accurate description of the level variation that is connected with motorization of population depending on the level of GDP. The function parameters allow changing the intensity of growth at different stages and the time of reaching the level of vehicle saturation, depending on the level of GDP per capita.

2.2 Energy consumption per unit of vehicle

The second important factor is to identify the laws of energy consumption dynamics per unit of a motor vehicle or specific energy consumption [8]. The research made by the authors has shown that the dynamic trend models allow a high degree of certainty to describe the patterns of change in the share of energy consumption in developed countries, showing relatively long downward trend of this indicator (over 20 years). At this, specific energy consumption per unit of GDP is described by an exponential dependence [9]. This means that with time, the growth of energy intensity tends to be decreasing evenly.

The duration of the energy consumption reduction dynamics in the transport sector is not as long as in economy on the whole. As a result, in addition to the exponential, energy consumption in the transport sector can be described by a number of other trends, in particular by a linear trend (Fig. 2). This is because the nonlinearity can be documented for a sufficiently long period of time. Therefore, when predicting specific energy consumption in the transport sector, the dynamics of energy consumption by a vehicle unit is assumed to be described by the same laws as changes in the energy intensity for the whole economy. The basic principles of open economy and participation in international trade allow us to extend these laws to the regions of the Russian Federation [10].

https://doi.org/10.1051/e3sconf/20198003010

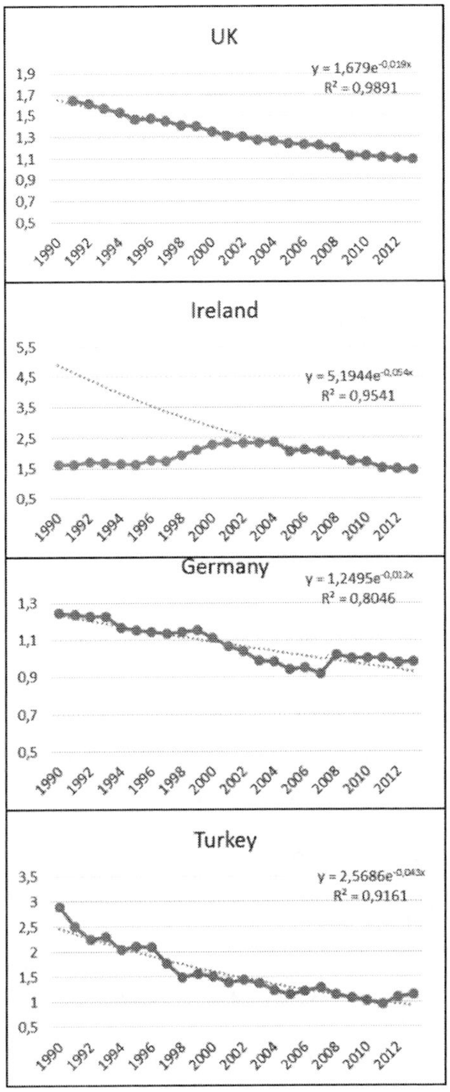

Fig 2. Dynamics of energy consumption of one vehicle in road transport, tonnes of oil equivalent per vehicle per yearSource: European Commission: Eurostat; World Bank

Most of motor vehicles are imported by Russia from Europe and Japan, which results in similarity of the energy consumption reducing processes in the transport sector, as well as European countries.

2.3 Changing structure of the energy balance for road transport

The role of alternative sources of energy (biofuels, natural gas, electricity) has been recently notably increasing in the transport sector, which is a limiting factor for an increase in demand for conventional fuels in the road sector – gasoline and diesel fuel [11,12].

For example, over the last three decades, the share of petroleum products was characterized by a continuous decline in the energy balance of Europe in the transport sector [13]. In the period from 1990 to 2005, the share of oil products in the region decreased from 98.7 to 97.2 %. The period of high oil prices and the economic crisis of 2008-2009 acted as incentive to switch to more economical alternative energy sources [14]. As a result, over 5 years – from 2005 to 2010– the share of oil products fell from 97.2 % to 93.5 %. By 2013, the consumption of gasoline and diesel fuel accounted for 92.9 % (Table 1).

Table 1. The structure of fuel consumption in road transport in Europe, %

The type of energy	1990	1995	2000	2005	2010	2011	2012	2013
Gasoline	57,8	53,7	47,2	38,1	30,5	29,5	28,5	27,7
Diesel fuel	40,9	44,9	51,2	59,0	63,0	63,6	64,1	65,2
Natural gas	1,2	1,2	1,4	1,7	2,1	2,2	2,3	2,5
Biofuels	0,0	0,1	0,2	1,1	4,4	4,6	5,0	4,6
Electricity	0,01	0,01	0,01	0,01	0,02	0,02	0,02	0,03

The world development in the last decade is characterized by the rapidly growing consumption of biofuels in the transport sector. So, in the period of 2005-2013, the share of biofuels in the structure of energy consumption in Europe has increased from 1.1 % to 4.6 %. The total transportation consumption of alternative energy sources such as natural gas, biofuels, electricity in 2013 was 7.1 %, whereas in 2000 their share was not greater than 1.7 %.

The share of diesel fuel tends to increase in the structure of energy consumption by transport is increasing rapidly. Diesel fuel is more economical in terms of fuel consumption (lower by 30-35 % versus gasoline). In addition, diesel fuel is cheaper in European countries due to lower tax rate. As a result, the use of diesel fuel as an energy carrier in the transport sector grew from 40,9 % to 65.2 % in the period 1990-2013.

Despite the increase in energy efficiency and the change in the structure of energy sources in road transport, it remains the largest and fastest growing source of greenhouse gas emissions. In general, in OECD countries, transport accounts for more than 30% of greenhouse gas emissions. In the period of 2013-2016 greenhouse gas emissions by fuel combustion in cars rose by 5.9% in the countries of the European Union (Fig. 3).

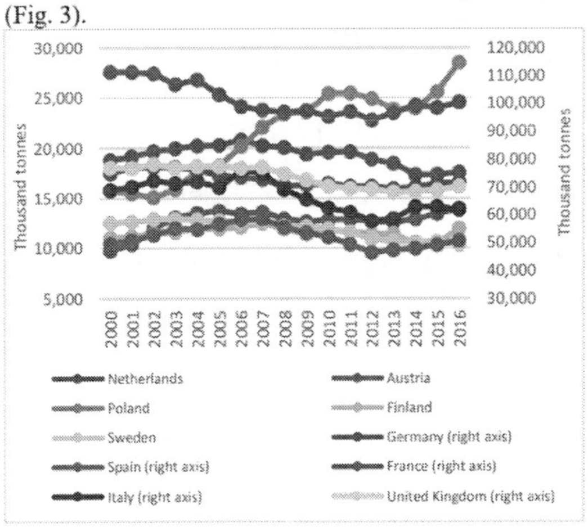

Fig 3. Greenhouse gas emissions by fuel combustion in carsSource: European Commission: Eurostat.

3 Methodology

3.1 Data

The pattern of energy consumption dynamics for road transport in Europe was analyzed on the basis of the Eurostat data presented for 32 countries for the period 1990-2013 [9].

For the analysis of changes in energy efficiency in transport, a database has been compiled for 32 European countries on such indicators as:

• Consumption of all types of energy in road transport;

• The total number of road vehicles, including cars, buses, motorcycles, trucks and tractors.

The construction of multifactorial regression and identification of key factors affecting the population's provision with motor transport was carried out on the basis of a database formed according to World Bank statistics. The database included 18 indicators for 63 countries. The selected indicators were divided into 6 groups of factors characterizing the comparative situation of countries:

• Ecological and nature-climatic factors;
• Cost factors;
• Infrastructural factors;
• Institutional factors;
• Technological factors;
• Socio-economic factors.

When calculating the predicted values of the amount of road transport for the period up to 2040, scenario forecasts of the world population (UN Department of Economic and Social Affairs) and the GDP forecast for PPP (World Bank) were used.

3.2 Empirical analysis

As was shown in the paper by Eder L., Nemov V. (2016), the cumulative forecast for energy consumption of road transport per capita was calculated as multiplying of the total number of vehicles by energy consumption of a vehicle unit (1) [4]. In addition, the population projections were used to calculate the absolute number of vehicles in the regions. They are based on the optimistic scenario of population dynamics in Russia according to the UN.

$$E_{tr\,it} = \left(V_{it}^{a} + V_{it}^{b} + V_{it}^{t}\right) * P_{it} * Y_{it} \quad (1)$$

$E_{tr\,it}$ – energy consumption of vehicles in region i at time t; V_{it}^{a} – the number of passenger vehicles per 1000 people in region i at time t (calculated by authors); V_{it}^{b} – number of buses per 1000 people in region i at time t (calculated by author); V_{it}^{t} – the number of freight vehicles per 1000 people in region i at time t (calculated by author); Yit – energy consumption for one vehicle in region i at time t (calculated by authors); Pit – predicted number of population in region i at time t (based on the forecast of the Department of economic and social Affairs of the United Nations) [10].

3.2.1 A model for predicting the number of vehicles

A model for predicting the number of vehicles is based on the patterns of change in the number of vehicles per person depending on the dynamics of GDP per person. Gompertz function best describes this functionality:

$$V_t = V_{max} e^{-\alpha e^{-\beta GDP_t}} \quad (2)$$

V_t – the number of motor vehicles per 1000 people at time t; V_{max} – "saturation level", desired equilibrium number of motor vehicles per 1000 people; GDP_t – GDP per capita in time t; α, β >0 – parameters.

To determine the specific factors that affect the "saturation level" of Vmax, a multifactor regression is constructed.

As a result of econometric analysis, the following 5 factors that have the greatest impact on the specific number of vehicles have been identified: CO_2 emissions per capita (E); percentage of urban population, % (U); cost of diesel fuel, USD / liter (P); hours per year spent on payment of basic types of taxes (H); density of roads, km per 1000 sq. km (R).

The regression equation has the form:

$$V = -163{,}72 + 14{,}06 * E + 2{,}3 * U + 133{,}83 * P - 0{,}16 * H + 0{,}04 * R \quad (3)$$

Inclusion of these factors in the model allows for calculation of the indicator V_{max} for each country, taking into account the specifics of countries: with different levels of economic development, economic structure, natural and climatic conditions, the level of development of institutions and infrastructure.

3.2.2 The model of energy consumption by unit of the vehicle

The model of energy consumption by a vehicle unit was based on time series analysis, with subsequent plotting the for dynamic trend model.

As shown in the previous section, the forecast values of consumption of energy in transport will be calculated on the basis of the exponential dependence:

$$Y_t = \beta * e^{-\gamma t} \quad (4)$$

Y_t – energy consumption of one vehicle in year t, o.e.; β – energy consumption of one vehicle in the initial period (period t=0); γ – coefficient of the intensity reduction of energy consumption.

Calculation of parameters of the exponential function for Russian regions was carried out on the basis of obtained regularities of the effect of consumption convergence of across regions. The authors have shown the existence of the effect of beta-convergence, showing the relationship of initial level and subsequent decline in energy consumption per unit vehicle. The effect of beta-convergence allows assessing the dependence of the rate of decline of energy consumption in the region i (γ_i) from its initial value (β_i):

$$\gamma_i = 0{,}219 * \beta_i - 0{,}126 \quad (5)$$

As the result, the higher is energy consumption per unit of a motor vehicle, the higher is the rate of its decline. However, when energy consumption by a motor vehicle approaches the average level of developed European countries, the rate of its decline becomes slower. To evaluate the effect of beta-convergence enables the use of trend models to predict energy consumption by road transport with a limited observation period for this indicator.

3.2.3 Emission of greenhouse gases by road transport

The forecast of greenhouse gas emissions is calculated on the basis of the predicted number of vehicles, the volume of demand for various types of fuel on motor vehicles and the assessment of the quality of environmental performance of road transport. The volume of greenhouse gas emissions in year t is calculated by the formula:

$$G_t = \sum_j E_{jt} * GF_{jt} \quad (6)$$

Where E_{jt} – the volume of demand for fuel j in year t; GF_{jt} – emission factor (specific emission of greenhouse gases per ton of burned fuel j in year t).

The volume of demand for the fuel j is calculated on the basis of the forecast of the amount of road transport, specific energy consumption and changes in the structure of energy carriers. The change in the GFjt emission factor over time is calculated on the basis of the average CO_2 emissions per km from new cars and the average life of the vehicles.

4 Results and discussion

4.1 Forecast of the number of vehicles

According to the built forecast of the number of passenger cars calculated according to the proposed model, by 2040 the growth of the number of cars by 13% is expected on the territory of Europe (Table 2).

Table 2. Forecast of the number of cars in Europe and Russia, million pieces

Country	2015	2020	2025	2030	2035	2040	2040/2015, %
Austria	4,7	4,8	4,9	5,0	5,1	5,2	111
Belgium	5,6	5,8	5,9	6,0	6,1	6,1	110
United Kingdom	29,6	31,0	32,0	32,8	33,4	34,0	115
Hungary	3,2	3,5	3,6	3,7	3,8	4,0	123
Germany	45,2	45,7	45,5	45,0	44,4	43,6	96
Greece	5,3	6,3	7,1	7,6	7,9	8,1	154
Denmark	2,4	2,6	2,7	2,9	3,0	3,1	126
Spain	22,7	23,7	24,0	24,2	24,3	24,4	107
Italy	37,7	38,8	39,2	39,4	39,4	39,4	104
Netherlands	8,0	8,3	8,5	8,6	8,6	8,6	107
Norway	2,6	2,8	2,9	3,0	3,1	3,2	124
Poland	20,9	22,5	22,9	23,0	22,7	22,3	107
Turkey	10,3	13,1	15,2	17,2	18,9	20,2	196
Finland	3,2	3,3	3,5	3,7	3,9	3,9	123
France	31,5	32,4	33,4	34,3	35,0	35,5	113
Sweden	4,6	4,8	5,0	5,2	5,3	5,5	119
Russia	44,3	50,9	55,4	57,8	58,8	59,2	134

The largest increase in number of cars is projected in Turkey (96%), Greece (54%), and Denmark (26%). In Germany, by 2020, the positive dynamics of the number of road vehicles will be replaced by a smooth decline, which is explained by the forecasted population decline. As a result, by 2040 the number of cars will decrease by 4% compared to the level of 2015. Russia has a significant potential for increasing the number of cars. Which is mainly due to the low current level of provision of the population with motor transport. As a result, by 2040, an increase in the number of cars by 34% is forecasted for Russia.

4.2 Forecast of energy consumption per unit of a motor vehicle

The calculations show that by 2040 the level of specific energy consumption in transport in Europe will be reduced by a factor of 1.5-2 (Table 3). In this case, there will be a significant reduction in the differentiation of this indicator among countries. As a result, the average level of energy consumption in Europe per car will be 0.75 toe.

Table 3. Forecast of energy consumption per unit of a motor vehicle, toe / vehicle

Country	2005	2010	2015	2020	2025	2030	2040
Austria	1,60	1,31	1,19	1,12	1,05	0,98	0,86
Belgium	1,38	1,36	1,19	1,08	0,99	0,90	0,75
Cyprus	1,36	1,24	1,07	0,97	0,89	0,81	0,67
Czech Republic	1,03	1,00	0,99	0,91	0,85	0,79	0,68
Denmark	1,60	1,55	1,44	1,34	1,24	1,15	0,99
Finland	-	1,08	0,95	0,88	0,82	0,76	0,66
France	1,09	1,02	0,95	0,89	0,84	0,78	0,69
Germany	0,94	1,00	0,93	0,87	0,82	0,77	0,69
Greece	1,15	0,81	0,73	0,69	0,65	0,62	0,55
Hungary	1,12	1,10	1,10	1,03	0,97	0,90	0,79
Ireland	2,05	1,70	1,46	1,37	1,29	1,22	1,08
Italy	0,88	0,75	0,71	0,65	0,59	0,54	0,45
Netherlands	1,27	1,18	1,14	1,07	1,01	0,95	0,84
Norway	1,43	1,41	1,26	1,16	1,08	0,99	0,85
Poland	0,74	0,77	0,63	0,61	0,59	0,57	0,53
Spain	1,17	0,97	0,87	0,82	0,77	0,72	0,64
Sweden	1,52	1,41	1,36	1,30	1,24	1,19	1,08
United Kingdom	1,24	1,13	1,04	0,95	0,86	0,79	0,65
Russia	-	-	1,33	1,24	1,16	1,08	0,94

4.3. Forecast of change of structure of consumption of energy in transport

Analysis of tendencies of changes in the structure of energy consumption in transport has allowed executing the forecast for the medium term. The share of alternative sources energy (LPG, natural gas, biofuels, electricity) in road transport in 2020 will increase to 9.7 %, in 2040 - 24 %. In the structure of energy consumption will be a significant reduction share of gasoline to 11.4 % by 2040. The share of diesel fuel will increase up to 2027 to the level of 69.1 %. However, an increase in the consumption of alternative energy sources will lead to stabilization and further reduction of the share of diesel fuel to the level of 64.6 % by 2040 (Table 4).

Table 4. Forecast of the structure of energy sources in motor transport, %

The type of energy	2015	2020	2025	2030	2035	2040
Petroleum products (gasoline, diesel fuel, LPG)	94,2	92,7	90,5	87,6	84,1	80,0
Other (biofuels, natural gas, electricity)	5,8	7,3	9,5	12,4	15,9	20,0

4.4 Estimation of greenhouse gas emissions

Based on forecasts of the amount of road transport, energy consumption and structure of energy sources, an estimation of greenhouse gas emissions from fuel combustion in road transport was carried out. Various factors have a multidirectional impact on the volume of greenhouse gas emissions. On one hand, it is the improvement of the environmental characteristics of engines, the transition to alternative energy sources; on the other hand, it is the change in the population and the growth of the vehicles number. Based on the obtained results, three groups of countries can be distinguished (Table 5).

Table 5. Forecast of greenhouse gas emissions by motor vehicles in Europe, million tons

Country	2010	2015	2020	2025	2030	2035	2040	2040/2015, %
Austria	12,8	13,4	13,3	12,6	11,9	11,2	10,4	77,7
Belgium	15,0	15,2	14,5	13,5	12,4	11,5	10,5	69,2
United Kingdom	70,3	68,9	67,8	63,6	59,2	54,9	50,8	73,7
Hungary	5,9	5,9	6,1	5,9	5,6	5,4	5,1	85,7
Germany	95,3	98,3	94,8	87,6	80,6	73,9	67,5	68,6
Greece	10,2	7,6	8,6	9,0	9,0	8,8	8,4	110,2
Denmark	6,7	6,9	6,9	6,6	6,4	6,1	5,8	84
Spain	51,9	49,1	48,7	45,7	42,7	39,8	37,0	75,3
Italy	62,4	62,9	59,7	54,9	50,2	45,6	41,4	65,9
Netherlands	19,6	17,3	17,1	16,1	15,1	14,1	13,0	75,2
Norway	5,6	5,5	5,4	5,1	4,9	4,6	4,3	78,7
Poland	25,4	25,6	27,3	26,2	24,8	23,2	21,5	84
Turkey	39,9	69,3	83,6	88,7	91,1	90,9	88,4	127,6
Finland	11,8	10,4	10,8	10,6	10,2	9,7	9,2	88
France	71,3	71,2	69,8	66,5	63,0	59,4	55,7	78,3
Sweden	11,9	10,7	10,4	10,0	9,7	9,2	8,8	82,9

The first group includes countries with a developed economy and a high level of provision of the population with motor transport. In these countries, the effect of improving the environmental characteristics of engines exceeds the effect of increasing the number of vehicles, which is expected to reduce greenhouse gas emissions by 2040. This group includes Germany, France, United Kingdom, Austria, the Netherlands and Belgium.

The second group of countries is expected to have a moderate reduction in greenhouse gas emissions. These countries have less potential to reduce emissions. This is due either to high rates of increase in number of road vehicles, or with a large share of alternative fuel at the present time. This group includes Hungary, Denmark and Poland.

For the third group of countries, further growth of greenhouse gas emissions is expected. The rapid growth in the number of road transport, as a result of both low current level of motor vehicle provision and high population growth rates, will lead to a significant increase in greenhouse gas emissions. This group include, first of all, to Turkey, where emissions are projected to increase by more than 27% by.

4.5 Estimation of greenhouse gas emissions in Russia

The rapid growth of greenhouse gas emissions from road transport in Russia is a result of the car boom. According to the Federal State Statistics Service, in 2010-2016 the number of passenger cars in Russia increased by 31.5%, exceeding 45.2 million units in 2016. As a result, the share of transport in the total volume of emissions of pollutants into the atmosphere in Russia increased to 45.1%. In 2017 emissions from cars in Russia exceeded 14,4 million tons, the leaders in terms of emissions are Moscow and the Moscow region. At the same time, per capita emissions in Moscow, as in St. Petersburg, are lower than the average Russian level (79 and 85 kg / person, respectively, versus 96 kg / person, on average in the country).

According to the forecast, the growth of greenhouse gas emissions from fuel combustion in road transport in Russia will continue until 2024 and reach the level of 15.4 million tons, which is 6.5% higher than the level of 2015. However, the gradual saturation of domestic demand, growth of provision of population with motor transport will lead to a reduction in demand for cars. As a result, after 2025, the reduction of greenhouse gas emissions is projected. In 2040, the volume of emissions will amount to 13.3 million tons.

Reduction of greenhouse gas emissions in the regions of Russia can be explained by various factors (Table 6).

Table 6. Forecast of greenhouse gas emissions by motor vehicles in regions of Russia*, million tons

Federal District	2012	2017	2020	2025	2030	2035	2040	2040/2017, %
Central	3,48	3,79	4,00	4,06	3,95	3,75	3,52	92,8
Northwest	1,25	1,40	1,48	1,50	1,47	1,39	1,31	93,4
South	1,13	1,65	1,51	1,53	1,49	1,41	1,32	79,9
North Caucasus	0,71	0,86	0,86	0,87	0,84	0,80	0,75	87,7
Volga	2,40	2,92	3,17	3,21	3,13	2,98	2,80	96
Ural	1,24	1,32	1,36	1,39	1,35	1,28	1,21	91,7
Siberian	1,80	1,81	2,02	2,05	2,00	1,90	1,79	99
Far East	0,68	0,70	0,76	0,77	0,74	0,69	0,63	89,9
Russia, total	12,68	14,45	15,15	15,38	14,97	14,19	13,33	92,2

* according to the Rosstat methodology, the data are not comparable with those of Eurostat

The largest region for the volume of greenhouse gas emissions is the Central Federal District, accounting for 26.2% of emissions. Despite the intensive growth in the number of vehicles in the district, by 2040 a significant reduction in greenhouse gas emissions is expected. This is due to the relatively faster pace of renewal of the fleet with the best environmental performance in the central regions of Russia. Reduction of emissions in the Far Eastern District is explained, first of all, by a decrease in the population. The least decrease in greenhouse gas emissions is projected in the Siberian Federal District.

As by 2040 the volume of emissions will be reduced only by 1%.

5 Conclusion

According to the forecast, in Europe the number of cars is expected to grow by 13%. The largest increase in cars is projected in Turkey (96%), Greece (54%), Denmark (26%). In Germany, by 2020, the positive dynamics of the number of road vehicles will be replaced by a smooth decline, which is due to the projected population decline. The calculations show that by 2040 the level of specific energy consumption in transport in Europe will be reduced by a factor of 1.5-2, while the share of oil products in the structure of consumed fuel on motor transport is reduced. Thus in Sweden by 2040, the share of oil products is projected to decline to 68%, in Denmark to 79%. On average, in 2040 the share of oil products in the structure of fuel consumption in road transport in Europe will be reduced to 80%. Nevertheless, in the countries of Europe, various trends in the field of greenhouse gas emissions are expected. Most countries have a high potential for reducing greenhouse gas emissions from fuel combustion on road vehicles. However, a rapid increase in the number of vehicles and low incentives for switching to alternative fuels will lead to the conservation of greenhouse gas emissions at the current level or to their growth. In Russia, emissions from vehicles are projected to reduce by 8% by 2040. The most progressive region is the Central Federal District, where the emission reduction is associated with the improvement of the qualitative characteristics of the motor vehicles in operation with a high rate of growth in the number of cars. However, in most of Russia's regions, emission reductions are associated with low growth rates or a decline in the population.

The research was supported by the grant No. 17-78-20218.

References

1. A.E. Kontorovich, L.V. Eder, I.V. Filimonova, M.V. Mishenin, V.Yu. Nemov, Oil industry of major historical centers of the Volga-Ural petroleum province: past, current state, and long-run prospects. Russian Geology and Geophysics, Volume 57, Issue 12, 2016.
2. A.G. Korzhubaev, I.V. Filimonova, L.V. Eder, I.A. Sokolova, Strategy of development of an oil-and-gas complex Siberia. Oil Industry, Issue 3, 2009.
3. U.Y. Arslangulov, Perspectives the global transport sector. "Energy" M .: Publishing House, 2009
4. L. V. Eder, V. Yu. Nemov, I.V. Filimonova, Prospects for Transport Energy Consumption: Methodological Approaches and Results of Forecasting. World of economics and management. Vol 16, N 1, 2016.
5. J.A. Paravantis, D.A. Georgakellos, Trends in energy consumption and carbon dioxide emissions of passenger cars and buses. Technological Forecasting and Social Change, Volume 74, Issue 5, 2007.
6. B. Liddle, S. Lung, The long-run causal relationship between transport energy consumption and GDP: Evidence from heterogeneous panel methods robust to cross-sectional dependence. Economics Letters, Vol.: 121, 2013.
7. J. Dargay, D. Gately, M. Sommer, Vehicle Ownership and Income Growth, Worldwide: 1960-2030. Energy Journal, Vol. 28, No. 4, 2007.
8. E.Y. Khrustalev, P.D. Ratner, Advances in technology and energy efficiency in industry and transport. The economic analysis: theory and practice, No 2, 2015.
9. L.V. Eder, Forecast of world energy consumption: methodological approaches, comparative evaluation. Mineral resources of Russia. Economics and management, No 6, 2013
10. O.V. Mazurova, Power consumption of the Russian transport on the background of global trends. Energy policy, No 3, 2010.
11. The Outlook for Energy of the world and Russia until 2040. M., ERI RAS, 2014.
12. A.F. Andreev, E.M. Karpel, Strategy of natural gas as a motor fuel in road transport. Problems of Economics and Management of oil and gas complex, No 3, 2012.
13. V.Y. Petrov, Passenger transport of the future: electric cars, hydrogen cars or traditional? Automotive industry, No 5, 2009.
14. A.V. Ivanov, Current State and trends of the world's use of natural gas as an alternative fuel on vehicles. Oil, gas and business, No 5, 2011.

Assessment of geochemical barriers at preservation of low-level radioactive waste storages

Anatoly Boguslavsky[1], *Olga* Gaskova[1, 2], and *Olga* Naymushina[1]

[1] Sobolev Institute of Geology and Mineralogy, Siberian Branch of Russian Academy of Sciences, 630090, Novosibirsk, Russia
[2] Novosibirsk State University, 630090, Novosibirsk, Russia

Abstract. The work considers geochemical aspects of the natural and man-made system for storing radioactive waste (RW) from one of the Siberian enterprises of the fuel and nuclear cycle. Careful geochemical testing of key sites of the system allows us to identify geochemical barriers that prevent the spread of uranium outside the sludge storage. In addition, experiments were conducted on the leaching of the sludge material in the laboratory using modern methods for determining the composition of solutions and solid phases, as well as experiments on uranium sorption on the main types of subsoil. Experimental and thermodynamic modeling of uranium deposition processes confirms that the system studied satisfactorily copes with the absorption of uranium taken out of the sludge storage due to dilution and sorption on rocks and bottom sediments. Particularly favorable are bottom sediments rich in organic matter, which bind uranium to organic-mineral complexes.

1 Introduction

The modern concept of handling low-level radioactive waste (LLRW) is to create a multi-barrier system that prevents the spread of radionuclides outside the storage facilities. Both natural and man-made structures can act as barriers to safety, which either detain pollutants due to low filtration rates or increased sorption characteristics [1-3]. With the preservation of sludge storage facilities that were designed dozens of years ago, the reliability of natural barriers is of particular importance [4-6]. Determination of reliability and sufficiency of the existing safety barriers in the future requires creation of a migration model for leading pollutants based on geological, hydrogeological and geochemical features of the site [7-9].

1.1 Object of investigation

The tail facility of OAO Novosibirsk Plant of Chemical Concentrates (NPCC) is located in several kilometers from city line of Novosibirsk.

Processing of ores and concentrates from 1954 to 1962 was carried out according to the acid-soda scheme and the waste products of the production were ore sands and a pulp of hydrated cakes. After 1962, the plant stopped processing ore, as raw materials there began to receive already purified and enriched uranium, devoid of the daughter elements, so carbonate uranyl became the main waste product.

The main masses of slurries of NPCCs are the tails of hydrated cakes. They are dominated by gypsum 60%, uranium content 0.13%, calcite, sulphates such as ettringite, bentorite, Fe oxides, amorphous phases, traces of sodium aluminate, chlorite are present. The individual mineral phases of uranium are absent. Uranium is in the form of isomorphic inclusions in carbonates and sulphates (including in gypsum). Samples of hydrate cakes of sandy and silt fractions (ore sand) contain primary grains of uraninite up to 0.5 mm in size.

The upper part of the geological section in the studied area is represented by overlapped Middle Quaternary formations. Directly on the site a unit of clay loams with a thickness of 3-9 m is distinguished by a weak water permeability (<0.001 m/day). They are the main water reservoir for groundwater in the area of the sludge storage. Above lie clay loams, interbedded with sandy clays and fine-grained sands. Their thickness is about 4 m and is characterized by increased water permeability (0.1 ÷ 0.3 m/day, up to 1.5 m/day). Sandy clays and sands are the most water-permeable soils in the area of tailings, and confine the main flow of groundwater.

Until creation of the tailings dam, the Pashensky ravine represented a deep, ragged ravine with a seasonal watercourse. After the creation of the technical dam, its thalweg is intensely swamped all the way to the Lake Kachimovskoye due to the increase in moisture stagnation and modern peat deposits filling underlying bed. In bottom formation of the ravine under the RW body, peat deposits were opened by drilling; their thickness in the thalweg is about one meter.

To determine the extent of migration and the redistribution of pollutants towards a distance from the sludge storage, approbation of surface and groundwater, bottom sediments and rocks of the technogenic system

NPCC - Lake Kachimovskoye - Lake Fonovoe and Lake Gniloe is carried out.

1.2 Methods

The determination of the elemental composition of solid samples was carried out by the XRF method at the elemental analysis station VEPP-3 of the Institute of Nuclear Physics of the SB RAS. Uranium forms were determined using an electronic scanning microscope Tescan MIRA 3 LMU. The mineral composition of the samples was defined by X-ray powder diffractometry on a DRON-3 diffractometer. The analysis of the elemental composition of solutions was carried out in the analytical center of IGM SB RAS using high-resolution inductively coupled plasma (ICP MS) mass spectrometry on an instrument manufactured by FINNIGAN MAT (Germany). The concentrations of anions were determined by titrimetric, turbidimetric, potentiometric methods. The species of uranium were determined by the sequential extraction procedure according to the scheme proposed in [10], which is a modified version of the Tessier scheme [11]. In this case, the water-soluble and exchangeable fractions are classified as readily mobile, while the carbonate is the nearest reserve, which is activated when the physico-chemical conditions and chemical compositions in the water/rock system (pH, pCO2, etc.) change.

Thermodynamic calculations in the heterophase 27-component H-O-C-Sl-N-S-Al-Si-Na-Ca-Mg-Mn-Fe-U-Pu-Am-Ru-Cs-So-Ni-Mo-Zr-Cu-Zn-Cd-Ba-Sr system were carried out at 25°C and a total pressure of 1 atm. by the GIBBS algorithm using the built-in thermodynamic bank UNITHERM of the HCh software package [12].

2 Results and discussion

2.1 Water chemical composition

According to chemical analyses of a surface water of the studied area, background concentration of chlorine (up to 2173 mg/L), sulfates (up to 1019 mg/L), sodium (up to 1418 mg/L), uranium (0.75 mg/L) are exceeded. However, in the process of removal from the sludge collector, the TDS and gradually decreases, pH value increases from 6.5 to 7.5. The general concentration of uranium also decreases to 0.0004 mg/L (Table 1).

2.2 Distribution of uranium species and sorption experiments

The finding of the sludge collector in the thalweg of the natural ravine causes the probability of flooding the storage with groundwater. The interaction of groundwater with radioactive waste consists of the following links: groundwater flow into the storage and interact with radioactive waste → interaction of contaminated water with peat → interaction of contaminated water with clay loam of eluvial cover → post-treatment of water in the permeable layer of sand and sandy loam.

To determine the scale of possible removal of uranium, its species in various rocks was determined (Table 2). The rocks were saturated with uranium and then subjected to leaching. It turned out that about 36% of uranium from peat passes into solution, but up to 60% is firmly retained in the solid phase (42% with carbonates and 16.9% directly with organics). In this case, the carbonate form predominates in all types of the boggy ravine, amounting to at least 40% (Table 3).

Table 2. Definition of uranium species in the main types of subsoil, %

Sequential extraction procedure	Sand	Clay loam soil	Peat	Clay
Water-soluble	2.9	1.3	35.7	19.1
Exchangeable	20.3	21.7	0.9	21.3
Carbonate	53.1	58.6	42.4	40.2
Bound to Fe/Mn oxides	1.7	1.6	2.3	1.1
Bound to organic matter	11.8	8.4	16.9	6.5
Residual	10.3	8.3	1.9	11.9

Table 3. The results of experiments on the sorption of uranyl-ion (a sample of sorbent 1 g, a solution of 50 ml)

$U_{initial}$, mg/L	Concentration in subsoil after sorption, ppm		Uranium sedimentation on subsoil, %	
	Clay loam soil	Sand	Clay loam soil	Sand
10	1469	164-311	74.9	8.4-15.5
1	142	22-23	74.7	11.9-12.3
0.1	18.8	5.5-7.9	98.9	28.3-51.9
0.01	2.1	0.46-0.8	95.5	48.1-81.9
0.001	0.15	0.05-0.06	87.2	67.6-84

Table 1. Chemical composition of surface waters with distance from the slime pit, mg/L

Site	pH	Cl^-	HCO_3^-	SO_4^{2-}	Conductivity, μSm/cm²	TDS, g/L	Na	Mg	Ca	Mn	Fe	Sr	U
At the output of the slime pit	6.5	2173	98	1018	8996	13.17	1418	152	1334	38	0.62	6.5	0.75
Kachimovskoe Lake	7.6	164	104	291.0	974.4	0.96	88	16	95	<0.02	<0.05	1.7	0.078
After Kachimovskoe Lake	8.0	137	137	245.0	907.2	0.87	76	16	90	<0.02	0.054	1.4	0.061
Gniloe Lake	7.8	64	98	94.5	464.9	0.39	34	12	41	<0.02	0.066	0.68	0.033
Fonovoe Lake (background)	7.8	12	366	1.6	471.8	0.32	7.8	12	70	0.008	0.14	0.55	0.0004
Ground water (background)	7.3	5	560	17	737	0.76	6.6	30	140	<0.002	0.05	1.27	0.002

A cover layer of natural clay loam is of great importance for preventing the spread of uranium (whereas in clays it turned out to be up to 19% of the water-soluble fraction and we assume the leaching of pore solutions containing uranium). Quartz, feldspars predominate in the mineral composition of clay loams. In the finely dispersed fraction, there are also mica, chlorite, calcite, a small amount of smectite-illite. According to the experimental data, in the concentration range of solution of 1 -2 mg/L, they can absorb uranium up to 0.01-0.02% (Table 3). Moreover, these concentrations are close to those found in underlying clay loams, which are saturated with uranium under natural conditions (up to 160 ppm).

The sorption characteristics of sands are much inferior to the initial clay loams; however, considering the extent of the sandy horizon, they play an important role in the overall purification. Most of the uranium, absorbed by the sands, passes into stable forms.

2.3 Water-rock interaction

In slurries of NPCC (tails of hydrated cakes), gypsum predominates 60%, uranium content reaches 0.13%. As calculations of the saturation indices show, the equilibrium between the liquid and solid phases in the slurry storage is not reached, the solutions have an unbalanced chemical composition (many anions). We believe that the reason is the preferential precipitation of cations in the form of carbonates and oxides in alkaline solutions, up to 70% of uranium is in the form of calcium uranate $CaUO_4$ (ORP ~ 0.3V). At the same time, slurry solutions are undersaturated with respect to gypsum and other complex sulfates, to all salts of chloride and nitrate ions. Anions should be taken out by drainage solutions in groundwater under the slime storage (model calculations showed that the precipitation of Cl^- and NO_3^- is possible only by evaporating the liquid pulp fraction more than a hundredfold). The ion exchange capacity of the rocks of the region is low; the main mechanism for removing Cl^- and NO_3^- is dilution with fresh solutions.

The calculated dynamic model of RW erosion by natural waters with the cessation of the intake of slimes indicates that after several tens of years the quick dissolving sulfates (gypsum, ettringite), then carbonates, fluorites will be washed out. After hundreds of years, only hydroxides and minerals of clays (chlorite, kaolinite, goethite) will be stable. The compound $CaUO_4$ is more stable than sulphates, so after gypsum and other soluble minerals leaching, even the relative enrichment of the solid phase with uranium will occur (due to a reduction in precipitation). Upon contact of mineralized drainage solutions with underlying low peat and clay or clay loam (bottom sediments), a number of processes of uranium immobilization are observed, including coprecipitation, chemisorption, physical sorption (ion exchange).

The data of thermodynamic modeling make it possible to differentiate the contribution of each of the processes. In particular, in clay loams and clays characterized by ion exchange, up to 20% of uranium is associated with the exchange fraction, carbonates accumulate up to 50% of uranium in the composition of solid solutions $Ca_xUO_{2(1-x)}CO_{3(solid)}$. Considering the duration of contact of these soils with polluted waters (about 60 years), one can speak of a stable state of the system. After passage of organic-mineral deposits, the water comes into contact with the groundwater of the area (background). The latter are nearly neutral (pH 7-8) and have a low TDS value and a hydrocarbonate-calcium composition. ORP and pH of contaminated and natural waters are close, therefore, when mixing solutions there are no barriers associated with oxidation-reduction and acid-base transitions. There is a gradual decrease in TDS and a change in the type of solutions from nitrate-chloride calcium to bicarbonate calcium only through the water/water exchange. But even in these waters there are no noticeable changes in the forms of uranium in the solution, the complex $Ca_2UO_2(CO_3)_3$ prevails (84.4%).

3 Conclusions

In the current geological and hydrogeological conditions, without additional measures for the waterproofing of structures, overlapping of slurries from above to divert meteoric waters is not enough to remove the technogenic deposit from the active water exchange zone. The intensity of uranium removal will be controlled by the amount of water interacting with radioactive waste. Given the reserve of mobile uranium, the duration of removal at existing migration rates is thousands of years. The system of geochemical barriers so far successfully copes with the spread of pollution into groundwater, but extrapolating the results for an extended period of time (n*100-n*1000) inevitably leads to the saturation of their buffer capacity, after which the rate of uranium proliferation will increase by orders of magnitude. In this case, there is a possibility of contamination of surface water. Therefore, under the current conditions, the only alternative to the disposal of radioactive waste from the sludge storage is groundwater abstraction. The most effective way for this is to arrange a barrage in the form of a sheet-pile bulkhead around the sludge pits, which will block the flow to the groundwater storage.

This study was financially supported the Russian Foundation for Basic Research, Project №17-05-00707 (*in situ* sampling and lab. chemical analysis, thermodynamic modelling) and by the Russian Science Foundation, Project №18-77-10029 (definition of uranium species in solids and sorption experiments).

References

1. Michalsen M.M., Goodman B.A., Kelly S.D., Kemner K.M., McKinley J.P., Stucki J.W., Jstok J.D. Environ. Sci. Technol. **40**, 7048–7053 (2006)
2. Carpenter, J., Bi Y., Hayes, K.F. Environ. Sci. Technol. **49**, 1078-1085(2015)
3. De Jong W.A., Apra E., Windus T.L., Nichols J.A., Harison R.J., Gutowski K.E., Dixon D.A. J. Phys. Chem. **110**, 11568–11577 (2005)

4. McKinley J.P., Zachara J.M., Liu C., Heald S.C., Prenitzer A.I., Kempshall B.W. Geochim. Cosmochim. Acta. **70**, 1873–1887 (2006)
5. Gaskova O.L., Boguslavsky A.E., Shemelina O.V. Appl. Geochem. **55**, 4, 152-159 (2015)
6. Gaskova Olga L., Boguslavsky Anatoly E. Procedia Earth and Planetary Science **7**, 288-291 (2013)
7. Boguslavskii, A.E., Gas'kova, O.L., Shemelina, O.V. Radiochemistry **58** 3, 323-328 (2016)
8. Langmuir D. Geochim. Cosmochim. Acta **42**, 547–569 (1978)
9. Murakami T., Ohnuki T., Isobe H., Sato T. Am. Mineral. **82**, 888–899 (1997)
10. Klemt E., Spasova Y., Zibold G., et al. Environmental Radioactivity in the Arctic and Antarctic. 67-70 (2002)
11. A. Tessier, P. G. C. Campbell, and M. Bisson. Anal. Chem., **51** 7, 844–851 (1979)
12. Shvarov, Yu.V. *Geochem. Int.* **46** 8, 834–839 (2008)

Morphological, optical and electrical properties of spray coated zinc ethyl xanthates for decomposition within a poly(3-hexylthiophene-2,5-diyl) matrix

Matthew Bishop[1, 2], *Bamidele* Akinwolemiwa [1, 2], *Lun* Zhang [2], *Chuang* Peng [3], *Di* Hu [1, 2, 4]

[1]International Doctoral Innovation Centre, University of Nottingham Ningbo China, Ningbo, 315100, PR China
[2]Department of Chemical and Environmental Engineering, and Energy Engineering Research Group, Faculty of Science and Engineering, The University of Nottingham Ningbo China, Ningbo 315100, PR China
[3]School of Resource and Environmental Science, Wuhan University, Wuhan 430072, Hubei, PR China
[4]New Materials Institute, The University of Nottingham Ningbo China, Ningbo 315100, PR China

Abstract. This work investigates the deposition of hybrid layers, for next generation in energy conversion, via spray coating. Understanding the effect that this deposition procedure has on these layers could lead to the rapid development of these technologies, for both laboratory applications and commercialisation. Synthesised zinc ethyl xanthate and poly(3-hexylthiophene-2,5-diyl) was spray-coated on substrates and heated to a temperature of 160 °C, to the hybrid film. Optical, morphological and conductive properties of these films were investigated and linked to the spray coating duration. It was revealed that shorter-duration spray times led to relatively low conductivity and smoother films, moreover, an increase in spraying duration also led to an increase in conductivity, but with increased roughness, from 6.178 nm to 8.317 nm. As the spray time was further increased factors, including film layering effects, led to a gradual decrease in conductivity accompanied by a decrease in the roughness. Smoother films were shown to result in higher light absorption, characterised by wider band gaps, which could be due to the crystal structure of the inorganic phase. The controllability of this rapid, facile, and inexpensive spray deposition process was then demonstrated in fabrication of prototype photovoltaic devices.

1 Introduction

Thin film metal chalcogenide/organic materials have been extensively studied in photovoltaics [1-5]. These active layers can be deposited through various methods, including spin coating [5-7] and spray coating [5, 7]. Within each of these deposition procedures there are many macroscopic parameters to be controlled, each of which alter structural features, such as thickness and roughness, of the deposited active layer. Spin coating deposition is a very popular technique, however, it is a very complex and time consuming technique [8, 9]. Spray coating is a very useful technique for high speed synthesis of active layers for laboratory applications while allowing for easy scale-up to commercialisation. The current understanding, however, of the growth of thin films during spray deposition is low, and it is important to better understand how the spray coating process can be modified to have a higher level of control on the morphology of the active layers to further improve the applicability of this technique [10].

Bulk Heterojunctions (BHJs) refer to multiple phases that are mixed within one active layer, and have been deposited using both spin and spray coating techniques [4, 5, 7, 11, 12], thus creating a three-dimensional blend of electron-donating and electron-accepting materials, which in turn offers a large surface area and can increase exciton dissociation [10]. Organic solar cells are highly can typically be produced at a lower cost than their inorganic counterparts, while being mechanically flexible with a comparatively lower negative impact on the environment, however, these organic systems usually have lower efficiencies and a reduced mechanical stability. BHJs comprising of both organic and inorganic components have been shown to combine their respective advantages, whilst synergistically reducing some of the disadvantages [13-18]. Therefore, it has been demonstrated that it is desirable to combine both organic and inorganic materials.

This work focuses on the synthesis of zinc sulphide, a low toxicity, wide band gap semiconductor [19, 20], via decomposition of zinc ethyl xanthate precursors (ZnXan), dispersed in a poly(3-hexylthiophene-2,5-diyl) (P3HT) matrix using spray coating deposition, without using a surfactant layer [21]. Using this single source precursor simplifies the method for producing these active layers, allowing for a one step deposition method [11]. The ZnXan precursor was selected due to the low temperatures required for decomposition to ZnS, causing little damage to the polymeric component, allowing for this single step deposition [22, 23]. The effects of varying the process conditions, namely spray time, was the main investigation undertaken in this work. This was conducted through studying how the variation of the spraying time influences the roughness and thickness of the films. Furthermore, the effects of the variations in the structural features on the resulting optical and electronic properties of the prepared films were also tested by using the films as active layers in prototype hybrid photovoltaic BHJ devices.

2 Experimental

2.1. Synthesis of ZnXan

The synthetic route followed a similar design proposed by Agrawal et al [11] for the synthesis of cadmium ethyl xanthate, simply replacing cadmium chloride with 0.7 g of zinc chloride. The off-white precipitate (ZnXan) was washed with de-ionized water, and dried at 50 °C.

2.2 Synthesis of hybrid BHJ

The mass of P3HT and ZnXan used was kept constant at 5.0 mg and 62.35 mg respectively, in order to obtain a one to one weight ratio of P3HT:ZnS. These were dissolved into a solution of 5.0 ml 1,2-dichlorobenzene (DCB) with 1.0 ml pyridine.

The spray distance and orientation was kept constant at a horizontal distance of 10 cm, between the nozzle and substrate, held perpendicular to the substrate. The spray time was the only parameter varied, using 10, 15, 20, 25 and 30 seconds (s) as variables. These deposited films were then heated to 160 °C for 30 minutes to decompose the xanthate.

These samples were then analysed for comparison, noting how changes in this deposition affected the properties of the films.

2.3 Characterisation techniques

The thickness and morphology of active layer surfaces were characterised in air using Veeco 3100 scanning probe microscope (SPM).
For ascertaining the optical spectra, the active material on the ITO-glass substrate was tested in a Cary 5000 UV-vis absorption spectrometer at a rate of 400 nm/s, between 350 – 750 nm. One cycle was taken for each sample.

The resistive properties of the films were characterised with the aid of AC Impedance Spectroscopy, using the PARSTAT 2000 electrochemical workstation from Princeton Advanced Instruments, at an amplitude of 20 mV, using a simple two-point probe method, without light simulation i.e. at dark open circuit voltage (OCV_{dark}), which was at a value of a few millivolts across all the films investigated. The bulk resistance was taken to be the real part of the impedance at 100 KHz

3 Results and discussion

It was proposed that three different factors could be altered from the deposition parameters:

1. Thin film growth mechanism,
2. Inorganic crystallisation,
3. Degree of order within the polymer.

Scanning probe microscopy (SPM) images of the ZnS:P3HT BHJs were collected at various deposition times. Table 1 shows a summary table of results collected.

Table 1. Summary table of the average results for all samples collected using SPM data.

	10 s	15 s	20 s	25 s	30 s
Thickness (nm)	36.58	47.82	65.87	49.31	39.41
Roughness root mean square (nm)	6.18	6.89	8.32	6.80	4.74

The surface roughness, see **Table 1**, shows that the longest deposition time displays the smoothest surface, and the middle deposition time displays the roughest surface.

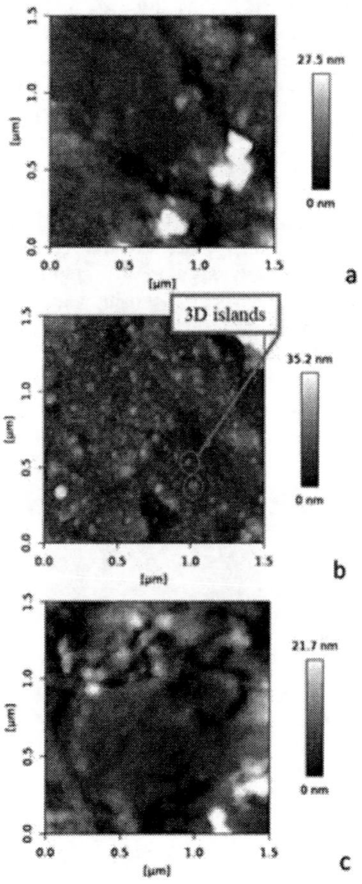

Fig 1. SPM for **a)** 10 seconds deposition, **b)** 20 seconds deposition, **c)** 30 seconds deposition.

The SPM images, see Fig 1, show the initial growth of a monolayer, before the formation of 3D islands, which in turn forms another monolayer, which suggests that a Stranski-Krastanov thin film growth mechanism might be ascribed to the formation of these films [40]. However, this mechanism does not fully describe the observed film growth as the longest deposition time decreases again in thickness. As previously mentioned, the spray orientation in this work is horizontal. Thus, with an increased amount of deposited precursor, the formation of macroscopic drips is possible, and this could reduce the film thickness whilst also lowering the measured roughness, this would potentially help with a potential photovoltaic device, increasing the Fill Factor as recombination sites are reduced. However, macroscopic drips were not noted during deposition and films remained homogeneous at all deposition times, suggesting another factor is the predominant mechanism in spray deposition.

Average absorption spectra were collected for the varying deposition times, and as the deposition time has an effect on the thickness these spectra were normalised by thickness, see Fig 2.

One key observation is that the spectra follow the same peaks, at roughly 370, 520, 550 and 600 nm, with the peak at roughly 600 nm being attributed to intra-chain interactions in P3HT [11]. This would suggest that the chemically active species within the active layers are the same, which is to be expected. The intensity of the peak related to P3HT intra-chain interactions, seen at roughly 600 nm, appears to be more intense for the middle deposition times, thereby suggesting a greater degree of order, which would result in an enhanced conductivity [25, 26].

Another key observation relates to the relative magnitude. As can be clearly seen in the graph the samples made from 30 seconds of deposition are the most absorbing. Samples made with 20 seconds of deposition are the least absorbing. This shows an inverse relationship between roughness, recorded from the SPM investigation, see Table 1, and the absorption. A potential reason for this trend in absorption could be related to the light scattering effects, i.e., rougher samples scatter light more, and if more light is scattered on the surface of the layer then less light will be absorbed.

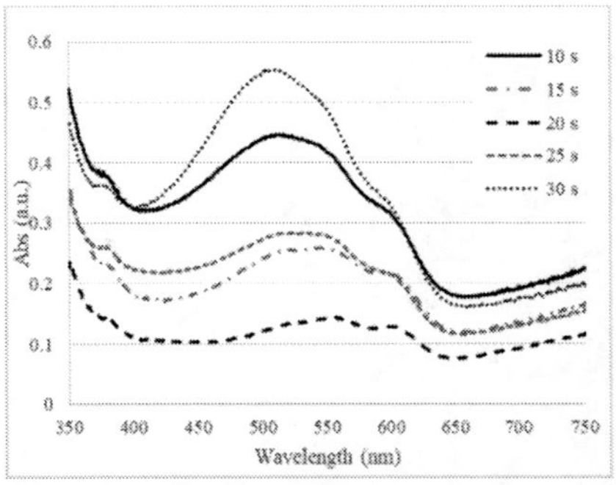

Fig 2. UV-vis absorbance spectra for the varying deposition times.

It was shown in previous works [6, 27] that band gap energies can be calculated from Tauc plots. The band gaps, see Table 1, were shown to mimic the same trend as noted from the intensity of the absorption spectra.

Table 2. Calculated band gap values from Tauc plots.

Deposition time (seconds)	*10*	*15*	*20*	*25*	*30*
Band gap (eV)	2.23	2.12	2.09	2.14	2.24

It has been shown that the lattice parameter is inversely proportional to the band gap [28], however, due to the nature of the active layer, x-ray diffraction (XRD) analysis could not be conducted. Since the active layer is primarily P3HT, with isolated nanocrystals of ZnS, effective spectra cannot be collected as the level of crystallinity is too low. Though, it is worth noting that the variation in band gap is quite small, and might not significantly alter the performance of the device.

As conjugated polymers use their macromolecular chains to facilitate the movement of delocalised charge. It has been well-established that both structural and morphological properties can contribute to the mobility of charge in conjugated polymers, particularly cross-links, loops, kinks and crystal structure [25, 26]. Therefore, it is expected that the morphological changes, induced through the variation in deposition time, would alter the electrical properties of the produced films. As seen in Fig 3, which subsumes the thickness, an obvious trend can be seen that clearly correlates the roughness with conductivity.

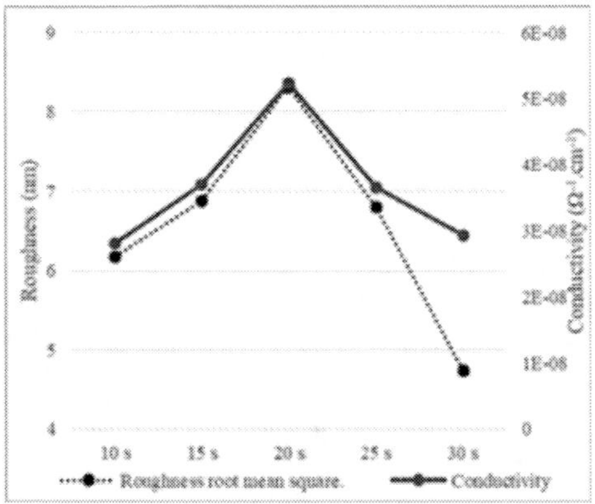

Fig 3. Correlation of roughness and conductivity with varying deposition time.

The degree of order in a polymer is expected to have varying effects on the conductivity of a film, since their electrical conductivity has been described through the so-called hopping mechanisms either within the backbone or across the interlinkage between macromolecules [29]. Therefore, higher conductivity is achievable with morphological that could allow for a relatively ordered macromolecular backbones. This obviously correlates with the surface roughness of the

https://doi.org/10.1051/e3sconf/20198003012

films due to the interrelationship between roughness and conductivity, see Fig 3. The conductivity is therefore indicative of how the polymer chain is orientated, which varies with varying spray deposition parameters. This shows that the deposition time has an effect on the formation of the nanostructure of the film, which in turn defines the degree of order for the P3HT deposited. Essentially, the varying degree of roughness could be correlated with the alignment of the P3HT chains, as a result of the film growth mechanism, which would affect the transport of charge within the active layer.

This could imply that the nucleation mechanism resulting in the film formation for spray-coating duration of 20 seconds provides energetic effects that are suitable for separation of the polymer chains, providing a more ordered structure, which is also reflected in the absorbance spectra, see Fig 2. Further increasing deposition time, and therefore time of nucleation, might result in the formation of detrimental cross-linkages, via ZnS nanoparticles, that could increase the transport path-length. Short duration spraying, on the other hand, would have insufficient time for the formation of separation, which could effectively lead to dead-ends [25].

4 Prototype photovoltaic device

In this section the potential application of this active layer in photovoltaic devices is demonstrated. These BHJs were spray-coated on ITO substrates, and then decomposed at 160 °C for 30 minutes. Finally, a gold electrode was deposited via sputter coating. The current-voltage (I-V) curves were calculated (measured using CHI 660C from CH Instruments) at ambient conditions, under one sun illumination (100 mW/cm^{-2}), key photovoltaic information is presented in Table 3 and I-V curves are presented in Fig 4.

Table 3. Photovoltaic feature of ITO-ZnS/P3HT-Au solar cells, with bulk heterojunctions deposited via spray coating at 10, 20 and 30 seconds.

	10 s	*20 s*	*30 s*
V_{OC} *(V)*	0.236	0.224	0.256
J_{SC} *(mA/cm^2)*	0.033	0.129	0.040
Fill Factor	0.622	0.495	0.637
PCE (%)	0.004831	0.01436	0.006552

Photovoltaic features of fabricated devices are reported in Table 3. As shown Jsc and the FF values all peak at 20 seconds of deposition, while the Voc and PCE is at the lowest value at 20 seconds of deposition. 20 seconds produced the highest Jsc observed due to favourable transport mechanism as confirmed by the electrical parameters, as demonstrated in Fig 3. Voc values stays at roughly 0.24 V for all deposition variations, which demonstrates little change to chemical characteristics with varying deposition parameter, as previously displayed in Fig 2, the minor variation is likely due to the varying band gaps, see Table 2, as these two characteristics have been shown to be interlinked. A significantly lower FF was observed for samples deposited at 20 seconds, indicating a diminished maximum power of the device. This is likely due to increased recombination of excitons at inorganic grain boundaries, attributed to the increase in surface roughness and noted 3D islands, see Fig 1. However, due to increased conductivity seen from films deposited at 20 seconds, see Fig 3, it was shown that devices deposited at 20 seconds vastly outperformed their counterparts. This suggests that the increased conductivity enhances the improved electron-hole transport in the device despite higher recombination.

From the features of the I-V curves, the maximum power point of the cells corresponds to the effect of the bulk resistivity and the absorption of the film. As was observed from Fig 4, the 20 s spraying time resulted in lower power compared to either 10 or 30 s, despite its conductivity been the highest. Therefore, the film morphology obtained at different spraying times is reflective of a combined connectivity between the active layer and the substrate, formed from the thin film growth mechanism and polymer chain order. Good connectivity between all the components would translate to higher power of the devices, however, the displayed power can be limited by the interconnection between the components of the active layer.

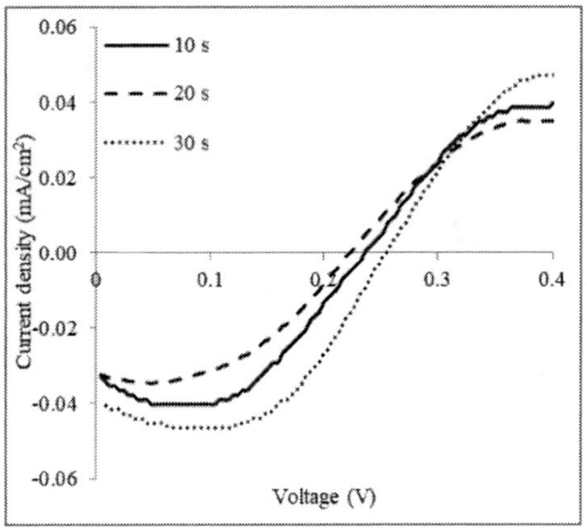

Fig 4. IV curves of ITO-ZnS/P3HT-Au with bulk heterojunctions deposited via spray coating at 10, 20 and 30 seconds.

The deposition time would, in practice, relate to the time required for various nucleation and combination process of the film to form on the substrate. Therefore, the measured photovoltaic properties reflect the ability of the spraying time to induce requisite power performance. It is suggested that the film formation process, is governed by a layering effect, which is a combination of thin film growth and polymer chain order, related to the time of spraying.

5 Conclusions

Structural, electrical and optical properties of in-situ synthesised ZnS:P3HT bulk heterojunctions prepared through spray coating were investigated in this article. The potential of the as-prepared films for photovoltaic applications were demonstrated, investigating how the deposition time and the effect this had on the microscopic features of the active layers. It was shown that short duration spray times lead to low order in P3HT chain, resulting in lower conductivity. As the spraying time increases there is an increased order of polymer backbones, allowing for the formation of optimal amounts of interlinkage and cross-linkage leading to the highest measured conductivity, while also increasing roughness and decreasing absorbance. As spray times increased further, additional factors, such as detrimental cross-linkage, came into play, resulting in lower conductivity films with decreased roughness and higher absorbance. It has also been shown that the smoother samples yield larger optical band gaps, which might be due to the crystal structure of the inorganic phase, though this could not be verified due to overwhelming amorphous features within deposited films. Photovoltaic devices assembled using active films prepared with these deposition parameters showed that as the conductivity of the materials increases the Jsc values also increase, however, the FF is dramatically reduced, which is suggested to be due to the increased recombination of excitons from inorganic grain boundaries. Furthermore, the observed surface roughness can also be linked to various layering effects which could be indicative of varying connectivity between the components of the deposited films.

This work received financial support from the Provincial Applied Research Programme for Commonweal Technology (2017C31104), and the National Natural Science Foundation of China (21503246). The author acknowledges the financial support from the International Doctoral Innovation Centre, Ningbo Education Bureau, Ningbo Science and Technology Bureau, and the University of Nottingham. This work was also supported by the UK Engineering and Physical Sciences Research Council [grant numbers EP/G037345/1 and EP/L016362/1].

References

1. M. Bredol, K. Matras, A. Szatkowski, J. Sanetra, A. Prodi-Schwab, Sol. Energy Mater. Sol. Cells **93** 662-666 (2009)
2. R.J. Zhou, J.G. Xue, ChemPhysChem, **13** 2471-2480 (2012)
3. S.L. Zhao, H.A. Wang, Y. Zhou, L. Liao, Y. Jiang, X. Yang, G.C. Chen, M. Lin, Y. Wang, H.L. Peng, Z.F. Liu, Nano Res. **8** 288-295 (2015)
4. N. Kumar, V. Dutta, J. Colloid Interface Sci., **434** 181-187 (2014)
5. E.A. Lewis, P.D. McNaughter, Z.J. Yin, Y.Q. Chen, J.R. Brent, S.A. Saah, J. Raftery, J.A.M. Awudza, M.A. Malik, P. O'Brien, S.J. Haigh, Chem. Mater. **27** 2127-2136 (2015)
6. A.B. Murphy, Sol. Energy Mater. Sol. Cells., **91** 1326-1337 (2007)
7. C. Girotto, B.P. Rand, J. Genoe, P. Heremans, Sol. Energy Mater. Sol. Cells., **93** 454-458 (2009)
8. Q.M. Zhang, M.J. Serpe, ACS Appl. Mater. Interfaces, **7** 27547-27553 (2015)
9. H.W. Hsu, C.L. Liu, Rsc Adv. **4** 30145-30149 (2014)
10. F. Aziz, A.F. Ismail, Mater. Sci. Semicond. Process., **39** 416-425 (2015)
11. V. Agrawal, K. Jain, L. Arora, S. Chand, J. Nanopart. Res., **15** 1697 (2013)
12. M. Tejos, B.G. Rolon, R. del Rio, G. Cabello, Mater. Sci. Semicond. Process., **11** 94-99 (2008)
13. M. Wright, A. Uddin, Sol. Energy Mater. Sol. Cells, **107** 87-111 (2012)
14. W. Jaimes, G. Alvarado-Tenorio, C. Martinez-Alonso, A. Quevedo-Lopez, H.L. Hu, M.E. Nicho, Mater. Sci. Semicond. Process., **37** 259-265 (2015)
15. N.A. Ogurtsov, A.A. Pud, O.P. Dimitriev, Y.P. Piryatinski, P.S. Smertenko, Y.V. Noskov, A.S. Kutsenko, Mol. Cryst. Liq. Cryst., **536** 33-40 (2011)
16. F.F. Cao, H. Wang, Z.H. Xia, X. Dai, S. Cong, C. Dong, B.Q. Sun, Y.H. Lou, Y.H. Sun, J. Zhao, G.F. Zou, Mater. Chem. Phys., 149 124-128 (2015)
17. J. Jung, Y.J. Yoon, M. He, Z.Q. Lin, J. Polym. Sci., Part B: Polym. Phys., **52** 1641-1660 (2014)
18. S. Obuchovsky, I. Deckman, M. Moshonov, T.S. Peretz, G. Ankonina, T.J. Savenije, G.L. Frey, J. Mater. Chem. C, **2** 8903-8910 (2014)
19. U. Jabeen, T. Adhikari, S.M. Shah, D. Pathak, T. Wagner, J.M. Nunzi, Eur. Phys. J. Appl. Phys., **78** (2017).
20. J.P. Borah, J. Barman, K.C. Sarma, Chalcogenide Lett., **5** 201-208 (2008)
21. J. Cusack, M.G.B. Drew, T.R. Spalding, Polyhedron., **23** 2315-2321 (2004)
22. H.C. Leventis, S.P. King, A. Sudlow, M.S. Hill, K.C. Molloy, S.A. Haque, Nano. Lett., **10** 1253-1258 (2010)
23. P.S. Nair, T. Radhakrishnan, N. Revaprasadu, G. Kolawole, P. O'Brien, J. Mater. Chem., **12** 2722-2725 (2002)
24. C. Ratsch, J.A. Venables, J. Vac. Sci. Technol. A, **21** 96-109 (2003)
25. U. Koldemir, S.R. Puniredd, M. Wagner, S. Tongay, T.D. McCarley, G.D. Kamenov, K. Müllen, W. Pisula, J.R. Reynolds, Macromolecules, **48** 6369-6377 (2015)
26. S.D. Oosterhout, M.M. Wienk, S.S. van Bavel, R. Thiedmann, L. Jan Anton Koster, J. Gilot, J. Loos, V. Schmidt, R.A.J. Janssen, Nat. Mater., **8** 818-824 (2009)
27. M.A. Buckingham, A.L. Catherall, M.S. Hill, A.L. Johnson, J.D. Parish, Cryst. Growth Des., **17** 907-912 (2017)
28. R. Dalven, Phys. Rev. B, **8** 6033-6034 (1973)
29. A.N. Aleshin, E.L. Alexandrova, Phys. Solid State, **50** 1978-1984 (2008)

AUTHOR INDEX

EDP Sciences – Web of Conferences
17, Avenue du Hoggar
Parc d'Activité de Courtabouf
BP 112
F-91944 Les Ulis Cedex A
France

ISBN 978-1-5108-7991-1

CURRAN ASSOCIATES INC.
proceedings
.com

4th International Conference on Advances in Materials, Machinery, Electronics (AMME 2021)

MATEC Web of Conferences Volume 353 (2021)

Wuhan, China
23 - 24 October 2021

edp sciences